Tourette Syndrome
Beyond the Unwelcome Companion

ISBN: 978-1-105-56240-2

Silver Run Publications

When I breathe, it breathes. When I speak, it speaks. When I try to sleep, it won't let me. Whatever I attempt to do, it's there...waiting to spoil the moment. To a doctor, it's a disorder, a medical oddity. To an onlooker, it's a spectacle - perhaps humorous, perhaps grotesque. To me, it's monster, a demon, a hellish beast who has no right to exist in my world or anyone else's – it's my unwelcome companion.

Tourette Syndrome
Beyond the Unwelcome Companion

Acknowledgments

Special Thanks to Sherry Joyce, Pat Priest, Dr. Neal Priest, Johnny Reynolds, Jennifer E. McKee

Disclaimer

Any use of medications or other treatments should be discussed in detail with a physician or a professional caregiver. The opinions and observations about Tourette syndrome in this book are solely the author's – mine -- and I make no claim to accurately describe anyone else's experiences. I am also not a medical professional and I am not offering any medical advice in this text. Tourette syndrome, like many other disorders, is manifested somewhat differently in every individual.

Table of Contents

Foreword

In 1996, I wrote The Unwelcome Companion, An Insiders View of Tourette Syndrome. It was published by Silver Run Publications. At the time, it was one of only a few books written by people who had Gilles de la Tourette syndrome (Tourette syndrome, TS) and were willing to talk publicly about it. Consequently, the book was widely read. However, the publishing company went out of business and the book went out of print in 2002. Used copies are still circulating quite actively on the Internet, and there is still much interest in the topic. I hope the book helped people with TS and their families to better understand this disorder.

Fortunately, awareness of Tourette syndrome has enormously increased in the past decade and things are getting a bit easier for those affected by this disorder. Scientists, physicians, psychologists, and others who study the workings of the brain have made giant leaps in the treatment of TS. Celebrities and spokespersons have greatly helped to spread both the awareness and the understanding of Tourette syndrome.

In spite of this wonderful progress, Tourette syndrome remains misunderstood by most and fully understood by no one. This enigma continues to drive parents, teachers, and caregivers in their quest for more knowledge about the disorder.

I greatly appreciate the efforts by professionals to uncover more about Tourette syndrome. I also feel that much more information needs to be shared by the people who actually have TS.

The book you are now reading contains a slightly updated version of The Unwelcome Companion followed by six new previously unpublished chapters. I hope you will find it interesting and that it will help readers get a better view into the mystery known as Tourette syndrome.

Part One: The Unwelcome Companion,
An Insider's View of Tourette Syndrome

Introduction

It was two o'clock on a Saturday morning. I looked out from the stage and saw the reflection of dim exit lights across a scuffed floor. The club's employees were beginning to clean tables and the last of the crowd was finally milling out the door. I was both tired and elated, for tonight's set had been a long but good one. There had been a sizable crowd and the band had played well, inspired by an enthusiastic audience.

As I put away my guitar and prepared to leave the stage, our vocalist walked up laughing. The club owner had just told her that he enjoyed the music but he was concerned about the guitar player's obvious drug problem. She giggled, "This guy thinks that you're some kind of addict in need of a fix. He claims he knows a doper when he sees one and the way you jerk around is a sure sign." She rolled her eyes. "I tried to explain but he wouldn't listen. He kept saying that your drug problem would do nothing but drag our band down."

My elation instantly evaporated. The band members may find these misunderstandings amusing but tonight's incident really bothered me. After years of dealing with similar negative reactions to my twitches and unusual movements, my patience was beginning to wear thin. The club owner was right about one thing; I do have a problem. But my odd behavior is not caused by illegal drug use. It results from a neurological disorder called Tourette syndrome (TS).

Though originally identified in the 1800's, this mysterious illness somehow slipped through the pages of medical journals until the late 1960's. Physicians and psychologists knew virtually nothing about TS, which was considered to be an extreme oddity.

Until recently, most people who had the disorder, characterized by involuntary movements and vocal outbursts, were thought to be insane. They were typically locked away in mental hospitals, given heavy doses of tranquilizing drugs, subjected to electroshock therapy, or in rare instances even given lobotomies.

Others with TS whose families came from superstitious backgrounds were thought to be possessed by demons. Some people with Tourette syndrome were subjected to ritualistic exorcisms. At one point, this was such a common occurrence that Tourette syndrome eventually came to be described by psychiatrists as "the medical cause of demonic possession."

Throughout history, TS has been both distorted by myth and inadequately defined in medical literature. It continues to be misunderstood today. Children with the disorder are sometimes punished for their bizarre actions, often perceived to be either mischievous or mentally disturbed. Adults are ostracized for their inappropriate behaviors and may experience difficulty holding jobs or developing successful social relationships.

Fortunately there has been increased discussion of TS in recent years and several excellent publications on the topic are now available. Most are written by physicians and researchers who have either treated or conducted studies of Tourette patients.

In spite of this wealth of information, many people, including some medical professionals, still have an inaccurate view of the internal mental processes that induce the behaviors associated with TS.

Such ambiguous perceptions can make it difficult for doctors to relate to and treat patients with this syndrome.

For this reason, I feel that it is important to present a view from within the patient's mind that provides the reader with a first-hand account of the nature and intensity of the battle that rages inside the Tourettic brain.

When I first decided to write this book, my biggest fear was that readers would think I was complaining or seeking sympathy. This is not the purpose of The Unwelcome Companion. I grew up with a wonderful family, and I have always had many caring friends. I have a moderate case of TS, allowing me to function reasonably well. I have little to complain about.

There are those who are not so lucky. Some are totally debilitated. Others have no love and support from the people around them. They must face both the torturous symptoms of their illness and the damaging curse of being rejected by those nearest to them. Still others live in communities where the doctors remain unfamiliar with TS. These patients may never discover the origin of their devastating symptoms.

In my travels, I have happened upon a few such unfortunate souls and have heard their stories. Some have been disowned by family members, fired from jobs, or expelled from schools. Many vaguely recall lost childhoods because their memories and dreams were destroyed by the long-term use of inappropriately prescribed medications.

Perhaps such injustices would not occur if there were a broader understanding of Tourette syndrome. Although there is no known cure for the disorder, in many cases proper medical care can drastically calm the symptoms. Understanding can accomplish even more by preventing much of the humiliation, rejection, and resulting emotional damage that can cause a person to become withdrawn, depressed, and unable to function in society.

The emotional and physical trauma of Tourette syndrome makes it one of the most disruptive disorders in existence. It upsets every aspect of a person's life, public and private, as well as the lives of one's peers, relatives, teachers, employers, and co-workers. They, too, must face the awkward task of dealing with the constant distraction, interruption, and embarrassment caused by tics.

People who have Tourette syndrome not only reveal their disorder; they unwillingly broadcast it. Onlookers watch helplessly as this curse twists and jerks its hostage into a contorted spectacle. The affliction makes its presence known almost as if it is proud, boastful of a nature that defies every civilized behavior known to humankind. It tries to consume the sufferer, exposing not the grace of the individual but the impediment of disease.

Chapter One: It Was Not Insanity!

Hospital waiting rooms sometimes seem brilliantly designed to intimidate visitors. The smell of antiseptic potions, the sound of muffled voices discussing life-and-death problems, and the fear of nearby airborne viruses can be extremely upsetting. As patients tentatively look around the room, they share similar thoughts. In addition to worrying about their own problems, they are curious about each other's diseases and the possibility of contracting them. A general uneasiness hangs in the air.

It was in such a place that I found myself waiting uncomfortably with dozens of other patients one summer afternoon in 1974. One by one we were called in to the examining rooms to be prodded and tested.

Some of us had already been through the probing and were anxiously awaiting the results of our examinations. At this point I had been in the hospital building for at least seven hours; I was ready to go home. Although my tests were not particularly painful, the wait was excruciating. It was not possible for me to sit quietly and patiently. I was constantly moving my hands and feet, clearing my throat, and shifting around in my seat in a futile attempt to get comfortable. However, my discomfort went far beyond the hardness of a waiting room chair.

I was never able to sit still even in the best of chairs. I could not rest in any position without having to endure the constant unwanted movements of my arms, legs, and facial muscles. I had difficulty speaking without repeatedly clearing my throat or grunting. True relaxation was something so remote that I had no recollection of how it could possibly feel. This "nervous problem" was the reason I was here in this waiting room, and at this point, it was coming on full throttle.

I was acutely aware of the stares of others as I fought the uncontrollable jerks of my neck and arms. The more I fought them, the more persistent and urgent they became. I felt like a freak on display.

As I continued my struggle against these contortions, I began to recall the years of endless battle with this condition. By now, I had seen at least a dozen doctors and had been tested and treated unsuccessfully for any number of diseases. My parents had reached their wit's end. This time, at their suggestion, I was having a complete physical examination at a hospital with impeccable credentials.

I had actually looked forward to this day, excited by the possibility of finally learning the nature of my problem. Maybe it could be treated. Maybe I could stop taking one ineffective prescription drug after another. I was tired of the side effects of these drugs and the withdrawal symptoms that resulted from switching medications every few months. Although I was barely in my twenties, I had already taken more prescription drugs than anyone I had ever known and had ingested more medications than many people consume in a lifetime.

It was a relief when at long last my name was called and I went in to see the team of doctors who had completed my extensive physical examination. I would finally know what was causing these strange symptoms.

There were five or six doctors and a couple of other medical professionals in the consultation room. They all seemed to look at me as if I had wasted their time.

"Son, we've checked everything we can think of," the chief doctor said. "There's nothing physically wrong with you. You just need to calm down."

In disbelief, I asked, "What causes me to jerk around? Why does my arm flip up? Why am I always blinking my eyes? Isn't there some reason for this?"

The doctor replied, "There is no physical cause; maybe you're just nervous. You should consider talking to a counselor. The problem is in your head, nothing more."

This verdict was crushing. I had truly hoped for some kind of answer this time. Was I really crazy? I did not think so yet the tests had shown no physical reason for my behavior. I began to get angry with the doctors for failing, frustrated with myself for not being able to control my behavior, and ashamed of being so messed up in the head.

For the next several years, I continued to search for another explanation. I felt that I was reasonably well adjusted, I had a clear understanding of reality, and I was able to think in a rational manner, even during my so-called "fits." In my opinion, that ruled out the possibility that the symptoms resulted from a severe psychological disturbance.

During my search, various doctors tried to figure out what was causing me to jerk around so strangely. No one could find a clue. I became increasingly depressed with every stranger's stare, every curious person's comment, and every doctor's failed diagnosis.

Finally, one day a friend mentioned a magazine article he had read that described a disorder of the central nervous system called Tourette syndrome. He felt that many of the symptoms noted in the article fit my condition. I had never heard of such an illness. Nevertheless, the motor symptoms he described sounded remarkably familiar, and I felt that it was worth checking into.

I made yet another medical appointment with my family doctor and discussed the possibility of my having Tourette syndrome. He was a dedicated, compassionate physician who seemed frustrated that, over the years, he had been unable to diagnose my condition. Like most family practitioners at the time, he knew nothing about Tourette syndrome. He referred me to a neurologist.

The neurologist immediately confirmed my suspicions. He went on to explain that Tourette syndrome is a neurological disorder in which the typical patient exhibits uncontrollable movements of various muscle groups and utters sounds or vocal outbursts. The disorder's severity and effect on the patient can range from mere aggravation to complete debilitation. In approximately one fourth of cases, the symptoms include coprolalia, the unintentional shouting of offensive or obscene statements.

Other manifestations of the disorder include the repeating of others' words or actions and obsessive-compulsive rituals (such as repeatedly touching people or objects and checking things over and over). Short attention span and hyperactivity are also common among the vast array of possibly related symptoms, as are learning disabilities, inappropriate or self-destructive behaviors, and depression.

At first I was shocked to learn that other people also had this disorder. I had always thought that I was the only person in the world who was fighting this battle. The fact that doctors had actually treated thousands of patients with similar symptoms seemed unbelievable to me. What a relief it was to learn that I was not alone, that this strange disorder had a name, and that it was not insanity!

Chapter Two: A Misunderstood Disorder

The first documented evidence of Tourette syndrome appeared in medical literature in 1825 -- although at the time, the disorder did not have a name. The patient in question was the Marquise de Dampierre, a French noblewoman. Her symptoms included most of the abnormal behaviors now associated with this disorder. She would violently move and jerk parts of her body, repeat statements or sounds made by others in an echoing fashion, and blurt out offensive and often profane phrases. In spite of her grotesque behavior, this woman was otherwise considered to be intelligent and sane.

Years later, a French neurologist, Georges Gilles de la Tourette, became fascinated by a disorder in which those afflicted exhibited uncontrollable "tics." In 1885, he studied and described to the medical community nine cases of this strange set of symptoms, or "syndrome." The doctor personally examined six of the patients including the Marquise de Dampierre, who was by this time an elderly woman.

In each case, symptoms included motor tics (involuntary gestures and movements of the extremities, torso, and facial muscles) and vocal tics such as grunting sounds, barks, yelps, and the occasional shouting of profanity. Many of the patients were troubled with obsessive thoughts and compulsive behavior patterns. Some would repeatedly perform grooming tasks while others touched objects over and over. In spite of these odd behaviors, all patients seemed mentally sound.

As a result of Dr. Gilles de la Tourette's studies, the disorder became known as Gilles de la Tourette syndrome, or the shorter version, Tourette syndrome. It is also referred to as Tourette's disorder, Tourette's syndrome, or simply TS.

Misunderstanding has plagued Tourette syndrome sufferers since long before the disorder had a name. In the past, people who made strange noises or bizarre gestures were usually thought to be either mentally unstable, completely insane, or possessed by demons. Although demonic possession seems an outlandish diagnosis today, how else could an eighteenth-century physician or clergyman have explained these outbursts of reprehensible remarks and odd gestures from otherwise sane people? These inappropriate, uncontrollable actions surely must have been the work of the devil.

It is now known that Tourette syndrome is a neurological disorder. It is not a psychosis and it is certainly not a manifestation of demonic possession. The condition is commonly inherited and is caused by neurochemical malfunction instead of either deep-rooted psychological problems or spiritual takeover.

Nevertheless, well-meaning but misinformed psychologists continue to erroneously assume that symptoms result from a patient's hidden hatred for a parent, suppressed memories of abuse, a desperate need for attention, or some other environmental stressor. Failed attempts at psychotherapy frequently cause feelings of confusion, guilt, and frustration, and they repeatedly prove to be of little or no value in curing tics. Fortunately, most mental health professionals educated within the past two decades are aware of TS and know how to recognize it. Sadly, it took until the mid-1980's for psychologists and physicians to become well informed about TS.

In many cases, a child with Tourette syndrome will start to show the first symptoms during a stressful period. This connection between stressful events and the onset of tics may cause others to mistake the child's symptoms for an emotional reaction to a tense situation. Circumstances may differ, but the scenario frequently involves a series of erroneous conclusions.

For instance, a child enters a new school. Shortly thereafter, he begins to blink his eyes rapidly and utter grunting sounds. These actions are initially viewed as attention-grabbing pranks. As the disorder takes its course, the symptoms worsen. The child, teased by other children and tormented by the internal tension of TS, begins to develop difficulty getting along with his classmates. He is soon branded as a problem child. His parents take him to a counselor who assumes that his behavior is nothing more than a psychological reaction to the stress of the new school. The child is in fact, beginning to show signs of an inevitable biological disorder that has nothing to do with the new school. Any stressful circumstance could have lowered the child's defenses, triggering an onslaught of his first symptoms of Tourette syndrome.

Unfortunately, similar situations sometimes result in the persecution of Tourettic children, not only by their peers, but by their parents and teachers as well. Due to widespread ignorance, people with tics frequently become outcasts or subjects of ridicule. Their bizarre gestures and sounds are thought to be intentional and are interpreted as deliberate acts of misbehavior. Under these circumstances, the disorder goes untreated and it is compounded by feelings of guilt and low self-esteem. Understanding on the part of parents, teachers, and physicians is an essential element in the management of TS. Only in recent years has the information to foster such understanding become available.

Medical reports published as late as 1982 declared that Tourette syndrome was "an extremely rare affliction." This is no longer considered to be true. Today, doctors estimate that as many as one in one hundred males and one in three hundred females has TS, approximately two million people in the United States alone. As more doctors are learning to recognize and diagnose TS, these estimates will likely increase. Available statistics also fail to reflect the vast number of individuals with mild symptoms who never seek medical attention and consequently remain undiagnosed.

One can only imagine the horror people with odd tics must have endured in years past. Untold numbers of fully cognizant, lucid people were either locked away in asylums or subjected to barbaric rituals, their lives destroyed by nothing more than a disruptive and misunderstood neurological disorder.

Chapter Three: Diagnosis of Tourette Syndrome

Other neurological disorders, including myoclonus, Huntington's disease, Sydenham's chorea, and Parkinson's disease may cause abnormal motor activity, jerks, tremors, or the loss of coordination. However, these symptoms bear little resemblance to the stereotyped tics of TS. "Stereotyped" tic movements and vocalizations are repeated identically as opposed to random twitches or tremors. This repetition is primarily what sets tic disorders apart.

It is estimated that over one fourth of all children may experience minor tics before adolescence such as excessive eye blinking or repeated facial grimacing. Common childhood tics may vary in location and severity and may last for years before disappearing. Because these bouts with tics are not permanent, they are referred to as Transient Tic Disorders.

Another fairly common childhood condition is Chronic Tic Disorder. A chronic tic remains relatively unchanged throughout its duration although it may also last for years. When more than one tic occurs, the condition is called Multiple Tic Disorder. If several tics are present and remain constant in location and nature, Chronic Multiple Tic Disorder is indicated.

Some researchers feel that childhood tic disorders may be related to Tourette syndrome. Others feel that they are actually mild forms of the syndrome. Unfortunately, there is no easy way to distinguish the appearance of a child's temporary tic disorder from the onset of TS. Tics commonly begin to appear in children at around age seven although they can show up as early as age three. If Tourette syndrome is indicated, it usually becomes evident before age fifteen. If both motor and vocal tics occur before age twenty-one and persist for more than one year, a diagnosis of Tourette syndrome is likely.

Tourette syndrome is a complex disorder that can mimic or coexist with other disturbances, sometimes making accurate diagnosis difficult. It is not unusual for a doctor seeing a TS patient for the first time to notice only the associated disorder, failing to recognize Tourette syndrome as a possible part of the problem. These coexisting disorders do not necessarily involve tics.

Hyperactivity literally means "over-activity" and is characterized by unusually high energy levels and excessive movement. A hyperactive child cannot sit still, is usually running or climbing, may have difficulty falling asleep, and is frequently a restless sleeper. Some of these symptoms may persist into adulthood.

Attention deficit disorder is a condition in which patients experience difficulty paying attention, are easily distracted, and frequently suffer learning problems. Impatience, excessive talking, inability to listen, and constant shifting from one activity to another are common in ADD.

Although most commonly seen in children, adults also suffer from hyperactivity, ADD, or both. Attention deficit hyperactivity disorder is a combination of the two. People with ADHD are usually fidgety and restless. They seem constantly on the go, often frantic and scattered. Quiet relaxation is generally impossible for these individuals; they must be moving about or talking constantly. ADHD sufferers are sometimes impulsive, irritable, and easily frustrated.

Varying degrees of ADHD are present in well over half of all TS patients, but most children who suffer from ADHD do not develop Tourette syndrome. It is difficult for a physician to predict TS in hyperactive children, as restlessness and other symptoms may be present for up to three years before any tics appear. In addition, ADHD is not necessarily related to TS.

Obsessive-compulsive disorder (OCD) is another condition that has been closely linked to TS. OCD can be debilitating and, like ADHD, can occur either independently or in conjunction with Tourette syndrome.

Obsessions are intrusive thoughts that constantly interrupt normal thinking patterns. These thoughts can sometimes be unpleasant or morbid in nature or they may simply consist of a phrase or mathematical process that plagues the mind. Obsessive-compulsive individuals are commonly driven by extreme perfectionism and constantly seek symmetry and order. A rearrangement of a bookshelf or a change in schedule may be unusually upsetting to a person with OCD.

Compulsions are often irresistible urges to perform actions and may be driven by obsessive thoughts. For instance, a compulsive need to wash one's hands might result from an obsession with the avoidance of germs and a feeling that one's hands are not yet thoroughly cleansed.

Compulsive activities are usually performed in a ritualistic fashion. Obsessive-compulsive individuals may feel that they must touch things a certain number of times, check and recheck light switches and door locks, or take an exact number of steps between two points. The need to carry out these rituals tends to be driven by an overwhelming feeling that dire consequences will result if these behaviors are not executed correctly.

Studies indicate that about half of all Tourette syndrome patients demonstrate obsessive-compulsive thought patterns and behaviors. These tendencies may develop in a more pronounced way well after the onset of tics. Specific tics may be performed in a mathematical sequence.

Although OCD is often listed as a possible symptom of TS, many neurological researchers feel that the reverse is true, placing Tourette syndrome under the umbrella of obsessive-compulsive disorders. Several studies refer to TS as a tic-related form of OCD. Tourette patients are generally compelled to perform tics, and OCD sufferers without TS tend to be driven by fears. In both cases, the person may feel that they must perform rituals aimed at alleviating these sensations.

In addition to ADHD and OCD, Tourette syndrome researchers have drawn links to numerous other disorders. These problems may be related to TS or they may simply be parallel conditions. Some may result from some of the same neurochemical disturbances found in TS. They seem to occur more often in Tourette patients than in the population as a whole although their exact relationship to TS is much debated. The list includes depression, phobias, panic attacks, dyslexia and other learning disabilities, aggressive or oppositional behavior, stuttering, and sleeping problems. Researchers have also suggested links to eating disorders and addictive tendencies in some Tourette syndrome patients. However, all of these connections are somewhat controversial and may not be truly related.

Although Tourette syndrome involves a variety of symptoms, motor and vocal tics continue to be the essential elements in the diagnosis of this disorder. Tics can be broken down according to type and complexity.

Motor tics are involuntary, stereotyped movements of an appendage, muscle group, or body part. These are usually rapid, abrupt gestures, such as jerking the head, slapping the side, contorting the face, or twisting the torso. They can be performed singularly or in sequence.

Vocal (phonic) tics include snorting, clearing the throat, sniffing, or in more severe cases, making animal-like noises, screaming, or blurting out phrases.

Both motor and vocal tics are divided into two classes: simple and complex. Simple tics usually involve only one muscle group or body part and have a very short duration. Stomping the foot is an example of a simple motor tic. A simple vocal tic is a brief noise made with the mouth, nose, or throat. Sniffs, grunts, throat clearing, clicking noises, and coughing noises are examples of simple vocal tics.

Complex motor tics consist of a combination of movements or gestures. They may be driven by obsessive urges and involve several muscle groups. Examples include retracing steps, jumping over cracks in the sidewalk, or repeatedly opening and closing a door.

Unusual complex motor tics include copropraxia (the making of obscene or offensive gestures) and echopraxia, also called echokinesis (the repeating or mimicking of others' movements). Coprographia is the writing of obscene words or statements. These unusual motor tics tend to appear primarily in the more severe cases of TS.

Complex vocal tics consist of repeated words, phrases, or noises. They often appear to be random or meaningless in nature or they may fall into one of the following categories:

Echolalia is the repeating of others' sounds, words, or phrases in an echoing fashion. A person with TS may pluck a random phrase from a conversation and rehearse it until the original speaker's inflections are perfected, eventually developing an uncanny ability to imitate others. Often mistaken for intentional mockery, this involuntary mimicry can be maddening for friends or family members.

Palilalia is the repeating of one's own words or phrases. Upon completing a sentence, a person repeats a word or phrase from that sentence several times until the tic has run its course.

Coprolalia is a vulgar or offensive vocal outburst. It ranges from the quiet utterance of a profane word to the shouting of offensive or obscene sentences. Sometimes an objectionable word or phrase is blurted out in mid-sentence, entirely out of context. At other times, an obscene remark may erupt from complete silence for no apparent, rational reason.

Though not necessarily vulgar, coprolalia is offensive. This embarrassing tic may also include the use of racist, sexist, or religious slurs, making it perhaps the most difficult social challenge facing the Tourette patient. The symptom is not easy to hide or disguise and is guaranteed to attract negative attention.

Because of its outrageous nature, coprolalia has received more press coverage than all other Tourette symptoms combined yet by most estimates it exists in less than one third of TS cases. This emphasis on the sensational has caused both the medical community and the public to incorrectly assume that coprolalia is an essential symptom of TS. It is highly probable that many cases of Tourette syndrome are overlooked because the symptoms do not include uncontrollable bursts of profanity.

Another type of pseudo-tic that is rarely discussed in literature is called the "mental tic." This irritating internal cycle involves no outward movement or vocalization. A repeated impulse stems from an obsessive thought and the need to resolve that thought. It may consist of an unpleasant idea and an immediate urge to counter that idea with a mental ritual, such as repeatedly counting to ten. Mental tics often involve mathematical processes and a fascination with symmetry.

Individuals differ, and therefore their Tourettic symptoms are not always the same. In some people with TS, an uncomfortable sensation may precede a motor or vocal tic, often related to a certain body part. For example, the mind may detect a buildup of energy or tension in a hand or foot.

A person may sense an imaginary object nearby and reach out to feel or move it, much like a mime touching an invisible wall. Although this type of phantom fixation may appear to observers to be triggered by a series of hallucinations, it is not. This action is driven by a vague yet convincing awareness that the phantom object, in some form, is there.

No two cases of TS are identical so no one person can speak for all Tourette syndrome patients.

A physician seeing a potential Tourette patient for the first time is faced with a challenge. There is no simple test for the slippery disorder's detection. In most cases, careful, repeated observation and questioning are the only diagnostic procedures used.

Under certain circumstances, the patient may be able to repress symptoms for brief periods of time. However, when the urge becomes unbearable or when the environment becomes more relaxed, the tics will emerge, frequently in an explosive manner. For this reason, the doctor may have to examine the patient several times. Candid video recordings are useful diagnostic aids as they enable doctors to witness behaviors normally concealed by the patient during an office visit.

Occasionally a physician may order an electro-encephalogram (EEG). This examination of electrical impulses in the brain has indicated abnormal activity in some TS patients. Another test, called positron emission tomography (PET scan), creates a visual image of the brain. Used more for research than diagnosis, this test has also shown unusual activity in some individuals with TS. As of this writing, neither PET scans nor EEG results provide conclusive evidence of the disorder.

Tourette syndrome is considered to be incurable yet treatable. The severity of symptoms may be affected by physical or mental stress, the presence of another illness, or any number of other factors.

Mild cases do not require treatment as long as the patient is able to function well without medication or therapy. Unless properly treated, more severe cases can destroy a person's quality of life. Deep brain stimulation (DBS), a surgery to implant an electronic regulating device in the brain, has had great success in treating severe cases of Tourette syndrome. However, DBS is a fairly new type of treatment and its long-term benefits are still being evaluated. As understanding broadens, perhaps earlier diagnosis, more effective treatment, and ultimately a cure are on the horizon for all who are touched by this multifaceted and complex disorder.

Chapter Four: A Tourette Story

Everyone who has experienced Tourette syndrome has a different story to tell. Tics and obsessions vary tremendously. Individual reactions to them vary as well. However, there are basic experiences that many of us with TS tend to have in common. All too often these include an unfortunate series of misunderstandings and fiascoes in the early stages of diagnosis and treatment.

Over a period of twenty years, I was tested and treated for everything from eye problems to manic depression. The possibility of Tourette syndrome was never mentioned; the cause of my odd behavior remained a mystery. Perhaps by telling my story, I can help others avoid the frustration of a similar ordeal. Fortunately, both doctors and the public are now drastically more aware of TS than they were when I was young.

As a child, I had boundless energy. I loved to run, climb, and stay constantly busy. Although this is normal childhood behavior, I was more than just a busy kid. My activity level was excessive. I was constantly fidgeting, jerking, and squirming, especially when eating or trying to relax. My parents became concerned and had me examined by a number of physicians. The diagnosis was always the same: "He's just hyperactive. Eventually he'll grow out of it."

As the years went by it became apparent that this hyperactivity was something more than a temporary childhood ailment. It was a preface to a much more complex problem. By the time I was twelve, my nervous energy had started to evolve; it seemed to be taking on a life of its own. Rather than disappear, my movements became more exaggerated.

My first real tics appeared in my early teens. I began to blink my eyes rapidly and repeatedly and started jerking my arm up and down by my side. By the time I was a sophomore, these movements were occurring every day, and they continued to increase in both frequency and complexity throughout my high school years.

As I developed, so did the tics. The blinking and arm jerks would dominate for a while and then a new set of gestures would take over, followed by another. I started jerking my head to the side and at times I would touch my nose repeatedly. Periodically, the old tics would return and combine with new ones, creating a virtual repertoire of unusual movements. Over the years, I learned to disguise a few tics but I was never able to hide them all. I had no idea why I moved this way. I just knew that I could not stop.

As a result of the tics, I began to develop problems in school, both socially and academically. Due to the twitching of my hands and arms, my handwriting was poor and erratic, often bringing complaints from teachers. I began to feel inferior and grew somewhat shy and reclusive because of my strange fidgeting. The constant struggle to conceal my symptoms was becoming more difficult. It was embarrassing and frustrating to have so little control over my body.

Due to the blinking, I was also erroneously prescribed eyeglasses for over a year as a child. This mistake caused quite a bit of problems. Finally, another doctor informed my concerned parents that I did not need vision correction at all. This year of unnecessary experimentation by the ophthalmologist amplified my learning problems because I could not see well and I suffered the ill effects of looking through incorrect lenses. It was a happy day for me when I could throw those glasses away.

I was lucky compared to most children with TS because my symptoms remained relatively minor until my late high school years. By this time, I had already established friendships and most of my classmates seemed to accept my odd nervous habits.

I was also fortunate with regard to vocal tics. The only ones I remember were an occasional clearing of the throat, snort, or grunt. I eventually learned to disguise most of them as allergy problems.

During this same period, obsessive-compulsive rituals began to develop. I recall counting the cracks in the school lunchroom ceiling tiles. At first, this felt like nothing more than a harmless game to occupy the time, but over the course of a few months, it grew increasingly necessary to count those cracks. Obsessions of this nature became more intense around my senior year of high school. I was rapidly becoming a slave to meaningless mathematical rituals.

The symptoms grew worse as I entered my early twenties. The struggle against tics and obsessive-compulsive rituals took every ounce of my time and energy. Fighting this battle started to feel like the sole purpose for my existence. At this point, I began an exhaustive search for answers.

Through it all, my parents did everything they could to find the proper help for my condition. At that time, such help was relatively unavailable. They were always supportive, and I often felt guilty for causing them to worry. I was totally baffled as to why I acted so strangely. I knew about the condition of hyperactivity but I had never heard that it could cause such bizarre symptoms. There had to be some other explanation.

By the time I reached my mid-twenties, I had been examined by several doctors and had been given a number of medications for a variety of possible disorders. I had been addicted to (and had withdrawn from) a couple of different tranquilizers. I was often prescribed very powerful sedatives because a restful night's sleep had become a rare luxury. Still, the tics continued, so my family doctor advised me to seek psychotherapy.

My first psychologist thought that he could cure me through hypnosis, but the therapy proved to be unsuccessful. I was never able to relax enough to allow myself to be hypnotized. We had several long sessions and he concluded that my twitches, jerks, and grunts were a physical manifestation of a deep psychological problem. After a few months, he diagnosed my condition as manic-depressive illness, and recommended several methods of treatment.

I followed the psychologist's instructions for improving my condition but nothing seemed to work. At his recommendation, I remained on tranquilizers and learned some relaxation techniques. I also read the self-help books he suggested. There are some helpful techniques to manage Tourette's offered now by psychologists; but at this time, they were not available.

Even with this combination of therapy and medication, I continued to fidget, jerk my arms and legs, and clear my throat continuously. Eventually, I began to doubt the psychologist's assumption that manic-depressive illness was the true source of my problems. I decided to go back to a general practitioner, searching elsewhere for the elusive cause of my strange behavior.

My next doctor insisted that I was hyperactive and that I should try a stimulant. He explained that stimulants had worked well in hyperactive children and they might be able to help a twenty-seven year-old. Although he said that this was an experimental treatment for someone my age, I was ready to try anything. He instructed me to immediately stop taking my tranquilizer (I had been on the same one for a year and a half by now) and he prescribed a drug called Ritalin instead.

For the next two weeks, I was in a daze. I could not eat or sleep and I went through a rough withdrawal from the tranquilizer, suffering from vomiting, headaches, confusion, and extreme depression. The jerking symptoms became unbearable. I finally had to stop the stimulant experiment for fear that it was beginning to cause serious harm.

I began to feel as if I had reached an all-time low. No one could figure out what was wrong with me, and every treatment that I tried seemed to make things worse. As I became more and more depressed, this doctor, having exhausted his repertoire of therapy options, recommended a psychiatrist. Once again, I had come full circle.

Although attempts at psychotherapy had failed in the past, I followed up on the suggestion and I had several sessions with a psychiatrist, hoping that his extensive training could help find some answers. After questioning me, he concluded that I suffered from self-induced stress, causing the tics and obsessions. He also thought that my excessive blinking resulted from eye problems and he referred me to an ophthalmologist. The eye doctor prescribed some drops to help reduce dryness but found no major problems. I tried the drops for months but they proved to be of no avail.

During the months of therapy, the psychiatrist taught me a few techniques for breaking self-destructive mental habits. He continued to insist that excessive worry, over-concern, and a tendency to focus on the negative were the roots of my problem. His methods helped my attitude but did nothing for my tics, so I moved on.

The guesswork continued for years. I was referred to specialist after specialist. The doctors checked for thyroid disorders, poisonings, diabetes, and numerous other diseases.

Perhaps the absence of profane outbursts prevented anyone from considering TS as my problem. After all, virtually every report published at that time focused on one Tourettic symptom: coprolalia. In retrospect, I can only speculate.

With each failed diagnostic attempt I became more discouraged. It is ironic that during this long period of trial and error, I probably did develop a few psychological problems. The long and unsuccessful search for the elusive skeleton in my mind's closet may have actually created one. I knew that something was very wrong and that its origin was likely to remain a mystery forever. This knowledge was an unyielding burden to my mind.

I was finally diagnosed with Tourette syndrome at age thirty two. Although I was not particularly pleased to have a strange neurological disorder, at least I had come to know the identity of my enemy, and this was a great relief.

Although it appeared that the battle was almost over, I soon realized that the struggle had just begun. It took years to discover the right combination of drugs for managing my tics and obsessions, and some of the prescribed medications created severe side effects, rendering me unable to function in everyday life.

Since that time, I have researched TS extensively in an effort to understand and cope with my condition. The sensations and emotions associated with it are interesting, varied, and complex. Words alone cannot adequately convey the mental activities endured with this disorder, for printed and spoken dialogue are limited forms of human communication. Although Tourette syndrome originates in the human brain, it embodies, and more importantly embraces, that which is inhuman.

Chapter Five: An Interruption of Regular Programming

A person with Tourette syndrome is forced to live two lives. He or she has one life dealing with the everyday stresses of health, career, relationships, and finances. This life is expected and considered to be normal for most people. But the other life is one of struggle, an existence dedicated entirely to battling an invisible enemy. While one portion of the mind maintains contact with the outside world, another is engaged in intense combat with invading Tourettic forces, fighting for the control of conscious thoughts and actions.

Using an arsenal of ammunition, the Tourettic foe attacks. One of its most effective weapons is the obsessive thought. An overwhelming notion is introduced, interrupting concentration, breaking down the rational mind's defenses, and allowing this intruder to quickly establish a state of mental mutiny. The Tourettic mind's prey becomes oddly convinced that a ritualistic sequence of tics must be carried out immediately. Sometimes within a fraction of a second, the initial obsession is followed by the enemy's second line of attack, an intense physical urge to tic.

In order for you, the reader, to understand this combination of sensations, imagine that you have a horrible itch and it is driving you crazy. You must scratch it immediately and repeatedly until it can no longer be felt. Until it is scratched sufficiently, you are overpowered by a feeling that disaster will surely come. The nature of the disaster is not always clear but the feeling of impending doom is without question.

When the consequences of this impending doom are known, they frequently involve your worst fears. The enemy may convince you that you will die, that your house will burn, or that a loved one's plane will crash if you fail to follow orders.

Whether vague or specific, the fear is just as real and overwhelming to the person with TS as anything in the universe yet it creates no actual terror or panic. These sensations are often not perceived as immediate threats but rather omens, urgent needs to take preventative measures. The rational, analytical mind is fully aware that there is no logical connection between moving a body part and preventing a disaster yet it is convinced that there may be a connection, one which was somehow overlooked. It is therefore an absolute necessity to tic and continue that tic until the tension relaxes and the sense of premonition dissipates.

An excellent description of this phenomenon can be found in the Hope Press book, "Don't Think About Monkeys." The following excerpt, reprinted with permission, dramatically relates the experiences of the father in a family of Tourette sufferers:

Excerpt:

It is like a dark, murky night that never ends; relief is brief and superficial, and then abruptly consumed. There is a sense of hopelessness that what his body feels is constant. The sensations are real, sometimes obscure, often frightening, never absent.

His body experiences a continuing series of peaks and valleys, but where, in other circumstances, the peak may mean elation or ecstasy and the valley tranquility, for the father the peak brings strain, confusion and pain, a sense than an unseen, unknown and unwelcome pressure is poking, pushing and shoving every part of his body.

It is closing doors time and time again until sensations inside him announce, "It is okay now, the door is closed correctly," even though he knew it was closed correctly the first, second, third time and every time after that. It was closed correctly even when the muscles in his hand and wrist contort in pain from grasping the door handle so many times until it is the "correct" time.

It is rolling up the car window a dozen times, locking doors over and over again, wringing out wet rags to the point where every muscle and bone in his hand aches. It is turning lights on and off dozens of times until he has halved the life expectancy of the bulb.

It is reading a sentence in a book or in a newspaper four, five, six or so many times that he has memorized it. But it must be done correctly, eyes beginning on the first letter of the first word, including every letter of every word as the eyes move from left to right to the next line and finally to the punctuation mark. "No, dammit, it's not right, do it again!" his body screams. Then he quickly glances at the flashing lights on the clock that tick away the seconds, adding another dimension to the compulsive ritual. A dimension unrequested, yet quick to volunteer, that coerces him into letting it be part of his body, to join the "fun." Then he reads the line again. It is still not right. His eyes jump to the clock. He is prohibited from seeing the actual flash but must focus on the clock between flashes. Then he returns to the page for round five of this sparring match with clock and book.

It is touching an object with his right hand and then forced by an eerie feeling of duress to touch the same object with his left hand, then repeating that ritual time after time until his body tells him,

"That's enough, I'm satisfied." He walks away, his energy and self-confidence diminished, then suddenly realizes that his body was only teasing him. It really was not satisfied. He returns to the same object and touches it again and again, first with the left hand, then with the right, or maybe twice with the left and twice with the right, or maybe...

Or perhaps his left hand does not wish to participate and he touches, touches with his right hand, unable to exhale until the need is satisfied. This time the body is not teasing, only temporarily inactive, a rest period that lasts for so little time. He walks away mumbling to himself, "Damn, why do I do it? Why can't I stop?"

End of Excerpt

These nonsensical thoughts and urges are difficult to understand because tics, obsessions, and their related mental activities are from a source other than the everyday, rational mind. Although tics seem to be senseless on the surface, they may make perfect sense and serve well-calculated purposes to the Tourettic part of the brain. It is as if the symptom-generating mind and the conscious mind operate independently yet simultaneously. Each has its own agenda. The intellect can generally maintain control but irrational obsessions will jump in at every opportunity, somewhat taking over until the tic is performed and the urge is resolved.

With complex, repetitive, or ritualistic tics, the process of resolution can last for several minutes. A person may repeat a tic much like a phonograph needle hanging on a scratched record. In contrast, with rapid or simple tics, the sensation, the tic, and the subsequent feeling of relief can all seemingly take place in a nanosecond.

A tic or ritual may also consist of not moving rather than moving. A person may feel the need to "lock in" on an object and stare at it until his or her eyes actually begin to lose focus, unable to move until the brain sends the command that it feels satisfied.

The relentless attack of these tics soon becomes exhausting and frustrating. The feeling that some unwanted outside power is possessing the mind and body is highly irritating. This takeover occurs much like a C.B. radio signal interrupting a radio or television broadcast. A radio tuned to a strong station is less likely to pick up interference than one that is tuned to a weaker signal. Likewise, when a person is extremely focused in work or play, the interference of tics and obsessions breaks in less often and with less intensity. During these focused moments, there is no need to fight the urge to tic because it is temporarily absent.

The mind's ability to squelch unwanted signals is virtually impossible to consciously achieve. It just happens. The relief is only temporary; it is inevitable that TS will break in with its relentless "interruption of regular programming" at some point.

I once had a job repairing copiers. Much of this work was fairly routine and I could perform it with little concentration. Therefore, it was extremely difficult to do my job efficiently due to the incessant, overriding signals of TS and the resulting motor tics. Periodically, my focus would keep the tics locked out but most of the time I was under "tic attack."

One day my tics were particularly bad and someone asked if I was being shocked by the machine. At the time, I was not in the mood for lengthy explanations so I replied, "It is just a little jolt, not too bad."

The tics sometimes caused me to break parts of the machine or bang my hands against sharp objects, creating aggravating small injuries. One symptom in particular appeared intermittently, severely interrupting the performance of my job. I would become unable to gently hold something in my hand and I would unwillingly squeeze a part or tool with great pressure. This resulted in my breaking an expensive circuit board on one machine. I finally decided to quit the copier repair business and focus fully on my primary occupation of playing music.

It may seem strange that a person with such a potentially embarrassing illness would choose to perform on stage in front of an audience. However, I have always been totally absorbed by music and determined that TS would not stop me from pursuing it as a career. I do feel that I would be a better musician were it not for the disorder because the tics can make quality practice difficult. During a performance however, the intense amount of concentration required to play music before an audience helps prevent the takeover of symptoms.

I have heard that others with TS experience the same phenomenon while they are heavily involved in a task, particularly one they enjoy. Perhaps it is this relief that has inspired many to choose music, theater, or sports as a career. Because these occupations require focus during a competition or performance, they provide excellent opportunities to minimize tics.

There are times when a person with TS appears to be controlling the tics even while he or she is not absorbed in a demanding activity. This is a different phenomenon than the remission of symptoms due to intense focus. It also goes beyond the ability to temporarily suppress tics and it is an actual escape from the grip of the disorder. On these infrequent occasions, the disorder seems to step back into the shadows and it is not felt at all. It is almost as if the illness temporarily disappears or perhaps enters a state of consciousness similar to sleep. To others, the absence of tics may seem to indicate that the person with TS has finally achieved self-control. This is an erroneous conclusion. The peaceful episode is merely a ceasefire, a break in the battle, one that is just as mysterious to the patient as it is to the observer.

Under normal conditions, the simplest, most fundamental tasks are sometimes the hardest actions to carry out for the person with TS. They require little focus and consequently provide no buffer against the invasion of tics. This gives the intruder carte blanche. Shifting gears in an automobile, shaving, and pouring a cup of coffee may seem like simple tasks to most people but they can become dangerous, near-impossible feats for those of us with tics.

A chore that I particularly dread is tuning my guitar, even with the convenience of an electronic tuner. Like every other guitarist, I pick up the guitar, pluck the E string, and reach for the first tuning key. My hand jerks back as if it has touched a hot iron. It then violently pounds against my side, sometimes causing deep bruises. While ignoring the pain, I reach for the tuning key again, managing to get near it this time. Suddenly, the guitar leaps several inches forward and then slams back into my lower stomach. I cringe for a moment, then continue.

Over the years, I have become well practiced at disregarding the physical pain inflicted by severe motor tics. Although I am fully aware of the discomfort, I am usually able to deny both the impact of and the reaction to the injury.

To the invading enemy this pain may be perceived as pleasurable, not in a sadistic way, but as verification of a successful attack. "It" seems to believe that physical pain caused by a violent tic indicates a victory, a hit, a winning score.

Ignoring the pain is relatively easy, but completing the desired task is extremely difficult. In the example above, I can't keep my fingers on the instrument's tuning keys long enough to make the necessary adjustments. With each attempt, I struggle to execute fluid and purposeful movements but my ability to do so is quickly overridden. I eventually tune each string with a slow series of jerky motions, sneaking in only a split second of adjustment at a time. Without tics, I could tune the guitar in about one minute, but with them it may take fifteen. To the Tourettic mind, this contest is a humorous game. To me, it is but one episode in a series of stressful experiences. It is also another of my many reasons to detest Tourette syndrome.

In order to understand how tics can feel, please imagine for a moment that Tourette syndrome is actually a form of demonic possession, as was once believed. Of course this is a silly notion, but a person with TS may often feel as if an unwanted entity lives inside his or her brain. This entity has a personality of its own and has set up shop in order to carry out its daily routine. This invader has needs, urges, feelings of satisfaction, a sense of purpose, clear awareness of task completion, and the will to take over. It knows what it wants and will not take "no" for an answer. It can compel a person to perform movements or vocalizations that would never be carried out otherwise.

If one resists, even manages to fight off the urge for a period of time, the entity will eventually win and make its victim pay dearly for the delay. The punishment is usually a much more severe explosion of the withheld tics.

When the person becomes fatigued from sickness, stress, or some other cause, the entity will seize the opportunity and have a field day. When the host's natural defenses are down, the invader can launch an unopposed attack, causing symptoms to appear in a more frequent, extreme manner. The Tourettic "demon" has a memory as well. It can store a multitude of tics and obsessions and recall them years later. Some tics will disappear for a period of time only to reappear with a vengeance as if they had never left. Tics and rituals continue to be added to the repertoire as this pseudo-parasite feeds from the memories and emotions of its victim.

The invader also sees things differently than the host and reacts accordingly. It will tap into the victim's knowledge of morals and etiquette and act in direct opposition to what it knows is proper conduct. In those with coprolalia, the entity will blurt out something offensive in an effort to humiliate the host into further submission.

Although coprolalia literally means "feces language," vocal outbursts of this type do not necessarily involve profanity. In one reported case, a TS patient often blurted out the word "hijack" when seated on an airplane. The entity observed the situation and knew exactly what to do in order to create the most havoc.

It may also harbor some jealousy of the host's ability to get along with other people. In an attempt to undermine this ability, it will conjure up a number of offensive behaviors aimed at sabotaging the host's relationships. The invader will try to repulse or aggravate others around its victim in order to fight its battle from the outside as well as the inside.

The demon can even force its hostage into self-harming behavior. I once had a tic attack so severe that I cracked several ribs by hitting myself in the side. This was obviously not an intentional act. I simply could not control the monster.

In the movie version of The Exorcist, one of the most disturbing scenes portrayed the possessed child stabbing herself with a crucifix while shouting profanity. Although this was a fictional story, this scene could easily have been a depiction of a severe Tourettic episode.

It is known that William Blatty, author of The Exorcist, studied a 1949 occurrence involving a young boy who was considered by the Catholic Church to be an actual victim of demonic possession. The church performed three rituals of exorcism in an attempt to help the child, whose symptoms included involuntary profane outbursts, growling, and violent, grotesque movements.

The final exorcism was thought to be successful yet the results of the church's follow-up examination are not well documented. Expert physicians later determined that the boy's unusual demeanor may have resulted from Tourette syndrome. They thought that the alleged success of the exorcism could be attributed to a spontaneous remission of tics, commonly found in TS. If such a mistake can occur in the mid-1900's, it is no wonder that witnesses in the distant past mistakenly viewed tics as manifestations of possession.

This illusion of being enslaved by demons actually helped lead to the discovery of Tourette syndrome. Dr. Gilles de la Tourette became interested in the numerous accounts of spiritual possession reported throughout Europe in the 1800's. He investigated several cases of demoniac behavior, and largely through these studies identified the syndrome which bears his name today.

I do not believe in possession by demons. Any references to demonic powers in this book are used for descriptive purposes only. It is my opinion that all alleged cases of demonic possession, if properly investigated, would turn up absolutely no evidence of the supernatural. Instead, these so-called "demons" would prove to be merely symptoms of a neurological, psychological, or physical illness. Unbelievably, reports of this nature continue. Imaginations run wild, witnesses exaggerate actual occurrences, and religious dogma has a strong suggestive influence on those who search for demons.

There are countless ways in which TS asserts its destructive powers. Over the years I have spent thousands of dollars on dental bills due to excessive grinding of the teeth. Others with TS have reported similar experiences. This activity may go unnoticed until sufficient damage to the teeth and gums brings it to a dentist's attention. Although many people grind their teeth during sleep, the Tourette patient may also grind and clamp teeth tightly together throughout the day, compounding the problem. Often a dentist will recommend a mouthpiece called a "night guard," in order to reduce damage from this destructive tendency.

Self-injury from TS may be much more blatant. People bruise and scratch themselves, chew their lips and tongues, and break bones in their hands and feet from tics. On rare occasions, extreme tics of the head cause retinal detachment or severe neck problems.

People with TS may also experience self-destructive urges. These are quite different from accidental injuries resulting from tics. They involve an obsessive thought that forces the person to deliberately perform a dangerous or painful act, such as closing one's eyes while driving or touching a hot stove. In the case of the latter, the burning pain acts as closure to the urge-relief ritual. To many researchers, it is unclear if such rituals are actually symptoms of Tourette syndrome or whether they stem from other, companion disorders. In fact, experts disagree on many aspects of this disorder.

Perhaps one of the most unclear and complex issues commonly debated concerns the volition of tics, whether they are actually voluntary or involuntary. Tics are generally described as involuntary and considered to be so by many researchers, yet some argue that tics might in fact be intentional executions of movement or sound.

The volition debate escalated when the Archives of General Psychiatry published a paper in 1980 detailing the experiences of a Tourette syndrome patient who recorded his observations of the sensations before, during, and after a tic. This patient, Joseph Bliss, concluded that his motor and vocal tics were voluntary acts performed for the purpose of satisfying unfulfilled urges. He stated that tics appeared to be involuntary to most witnesses but actually were not. To the Tourettic mind, these acts are intentional.

Some observers assume that Tourette syndrome is caused by the random misfiring of neurons in the brain and consequent reaction of the muscles affected. Tics are not that simple. Unlike convulsions, twitches, or spasms that are usually merely observed by the conscious mind either during the event or after the fact, tics are purposeful in design, stereotyped, and definitely deliberate.

I have been asked, "Why does your arm move like that?" It is difficult to explain, but my arm does not simply move. A chain of command is executed throughout the body's neuromuscular system, and the arm is ordered to move. However, the commander is not me, at least not the me I recognize as a conscious person. The tic must be performed in the correct, intentional manner necessary to relieve the urge and satisfy the master, which in this case is the Tourettic part of the mind.

TS is so powerful that it can almost completely take control over one's body, in some cases forcing a person to withdraw from society. Severely affected people with Tourette syndrome may go for months without leaving home due to embarrassment, shame, and public aversion to their strange behavior.

Generally, the only way doctors have been able to calm this monster is to drug it. Unfortunately, to drug the demon is also to drug the person in which it resides. The enemy can never be destroyed because it is a part of the patient. It knows every thought that enters the person's mind, and it cannot be tricked. These factors make it difficult to subdue the disorder without negatively affecting the conscious mind of the patient.

During a radio interview, I was asked to point out the differences between Tourette syndrome and Multiple Personality Disorder (MPD), a condition in which the patient displays two or more distinctly individual personalities. The reporter pointed out that both disorders seem to have the power to "possess" the sufferer. This observation is correct, but the two disorders are not related at all.

Tourette syndrome is an inherited condition, and its outward symptoms are induced by chemical imbalances in the brain. The primary psychological effects of TS are the anxiety and depression that result from dealing with the biological disorder.

Multiple Personality Disorder is considered to be a psychologically based condition in which the patient develops different personalities in order to cope with emotional trauma, such as abuse as a child. Because an enormously high percentage of MPD cases involve patients who have known histories of upsetting experiences, this rationale is thought to be accurate.

Another significant difference between the two disorders lies in the patient's awareness of inner conflict. A person with MPD may be unaware that other personalities occupy the same body. There is no rapid, relentless tug-of-war for control. Either one personality rules or another. MPD is a disassociative disorder in which a person loses association with his or her primary personality.

Tourette syndrome is completely different. The patient is totally aware that the mind is engaged in a never-ending struggle for control. The tics, obsessions, and compulsions may seem to dominate briefly but the person is always mindful of the situation and feels a dire need to recover command. Tourette syndrome does not cause a person to lose awareness of reality.

This awareness of reality is both a blessing and a curse. It is a blessing in that a person with TS remains lucid and can usually gain control in a matter of seconds, before the intrusive entity fully carries out its often reprehensible and destructive behavior. However, it is a horrible experience to witness oneself uncontrollably performing contortions, vocalizations, and gestures even for a moment, while being fully cognizant of their grotesque nature.

Chapter Six: Causes of Tourette Syndrome

The human brain is an electrochemical mechanism. Its chemical makeup, operating in concert with electrical impulses, helps to form our personalities. A change in this chemistry can cause a change in behavior. Imbalances within the brain are responsible for a myriad of disorders, including Tourette syndrome. If a human can be regarded as an electro-chemical being, TS may be considered an entity of the same nature, only one born of malfunction.

Utilizing pedigree studies of the families of patients, many scientists have concluded that Tourette syndrome is an inherited disorder. A few researchers question the certainty of this conclusion. In about one-fourth of Tourette patients investigated, pedigree studies indicate no apparent genetic explanation for the transference of the disorder. One school of thought suggests that these may be spontaneous cases related to injury or difficulties during birth. Most geneticists maintain, however, that the disorder is inherited despite the fact that it is sometimes difficult to find a relative with symptoms.

The main culprits in the manifestation of Tourette syndrome are believed to be various neurotransmitters, the chemicals that carry signals in the brain. The principal neurotransmitters thought to be involved are a group called biogenic amines, including serotonin, dopamine, and norepinephrine. One important factor in Tourette syndrome's influence on behavior is its ability to cause a "disinhibiting" effect. When a tic comes on strong, an individual may perform acts or make statements that would otherwise be censored by his or her conscious mind.

Serotonin is a neurotransmitter believed to be involved in inhibition. Both human and animal studies have proven that an imbalance of this chemical can cause drastic behavioral changes, leading to aggression, obsessive-compulsive thinking, depression, addictive tendencies, and other conditions. The brain of a person with Tourette syndrome may not adequately process serotonin, resulting in improperly governed thought processes.

Dopamine, another neurotransmitter, affects muscle movement as well as behavior. A deficiency of dopamine has been found in patients suffering from Parkinson's disease, a condition characterized by difficulty in movement execution, slow or rigid motion, and tremor. Increased dopamine levels can cause exaggerated behaviors including aggression and increased sexual activity. Overactive dopamine-processing neurons or excessive dopamine levels in certain parts of the brain may also contribute to the rapid tic movements found in TS patients.

Dopamine processing is regulated by another neurotransmitter called norepinephrine (also called noradrenalin). It modulates the actions of many of the brain's other chemicals. An imbalance of norepinephrine upsets overall neurochemical harmony, allowing the onset of a number of symptoms.

Many of the chemical imbalances affect the structures of the brain known as the basal ganglia. These centers, located deep within the brain, act as relay stations for neurochemical messages. Imbalances can also affect the limbic system, the part of the brain controlling emotions. A disturbance within this system can cause decreased inhibitions and inappropriate reactions. Many chemicals in the brain can influence human behavior, and the number contributing to TS is still unclear.

My purpose in writing this book is to explain one person's view of how the disorder feels. In speaking with others with TS, I believe that many of my experiences with tics and other symptoms are quite common. I am certainly not qualified to discuss in detail the neurochemical (and other) processes that cause these symptoms. I will leave that part up to the trained researchers, and I thank them for their efforts to understand this disorder. Their ongoing study of the brain will undoubtedly turn up additional clues for solving the complex riddle that is Tourette syndrome.

Chapter Seven: The Drug Dilemma

Discovering the nature of a common illness and prescribing the appropriate medication is usually a relatively routine matter for a trained physician. Neurological disorders, however, can be much more difficult to properly diagnose and treat. The complexity of these disorders, the myriad of treatment options, and the potential adverse effects of medication must all be considered.

Before I was diagnosed with Tourette syndrome, I was prescribed a number of drugs that not only failed to help my symptoms but caused severely annoying side effects. I soon grew tired of the headaches, confusion, exhaustion, blurred vision, and other problems associated with these drugs. Memories of those years now seem vague. Much was lost in the fog of inappropriate medication. I cannot remember some periods of my life nor can I recall the name of every drug that I was prescribed. Many of these pharmaceuticals apparently acted as chemical sledgehammers, temporarily pounding every ounce of cognitive ability from my brain.

In spite of this, I was determined to cooperate with my doctors in an effort to find a solution to my problem. In retrospect, I realize that some of these physicians apparently had no clue as to what my illness could be yet most seemed unwilling to admit it. They were merely shooting in the dark.

After being correctly diagnosed, I thought my years of living at the mercy of unpleasant side effects were over. I was wrong. It took six additional years of experimentation before my physicians and I were able to find a proper combination and dosage of medications, one which reduced the symptoms without destroying my ability to function.

I was prescribed a major tranquilizer called Haldol and tried various dosages of it for years. I assumed that it was the only drug available for the treatment of TS. When I mentioned problems with side effects to my neurologist, a change in dosage was the only suggested remedy. My intolerance for this drug, combined with my lack of knowledge, resulted in years of unnecessary anguish. Finally, I began to research the subject. I also changed physicians.

Once I became aware that other medications were available, I started working, with my new doctor's help, to find a suitable treatment. Although experimentation was sometimes rough, even the most intolerable of drugs provided some escape from the relentless attack of tics. By this time, I had learned to accept the side effects as simply part of the quest.

In some cases, particularly with children, Tourette syndrome patients seem to be medicated primarily for the benefit of those around them. Once a person's tics are brought under control, those nearby feel relieved by the peace and quiet but the patient may be stuck in a private nightmare. The true challenge for the physician is not only to control the disorder but to preserve or improve the patient's quality of life in the process.

For a person with TS, the first and most important step is to find a doctor who is either knowledgeable about the available treatment options or is willing to thoroughly research the topic. Physicians with a clear understanding of TS are aware that treatment of this complex disorder requires considerable involvement on the part of the patient. No blanket approach will work. If one medication is intolerable and another exists, the patient should be informed.

The patient has equal responsibility in the process. Ineffective drug therapy is not always the doctor's fault. If a medication is causing unpleasant side effects or if it is failing to reduce symptoms, the patient must speak up. Without this dialogue, even the best of physicians cannot sufficiently treat the disorder.

It can be quite a challenge for any physician to manage just one complex illness, such as obsessive-compulsive disorder; but Tourette syndrome is particularly difficult to treat. The presence of so many symptoms and possibly related or coexisting conditions can create a neurological and psychological Medusa. One drug may alleviate a certain symptom but leave the others untouched. In some cases it can even make them worse. A patient can become frustrated and fail to give the doctor adequate time to complete a necessary medication trial. Some drugs take months to reach their full effectiveness. Instant relief without adverse side effects is extremely rare.

With or without treatment of any kind, this disorder tends to change in both severity and nature. Throughout the years, its tendency to wax and wane has caused many a patient, physician, and psychologist to misjudge the effectiveness of various therapeutic methods.

Outside influences also have an enormous effect on TS. Pressure in the workplace, stress at home, or any uncomfortable situation may trigger an increase in both tics and obsessive-compulsive behaviors. Sleeping too much or too little, excitement, anticipation, and other emotional stimuli can drastically affect symptoms.

Any number of foods, drinks, and medications can have an effect on TS as well. Over-the-counter cold remedies and caffeine are good examples. These substances can aggravate symptoms, which may lead to an incorrect evaluation of the patient's medication. Diet, drinking habits, and use of over-the-counter medications should be discussed with a physician in order to help minimize confusion.

There are no perfect treatments for Tourette syndrome, yet given the wide array of available options, most patients no longer have to accept the choice between an intolerable treatment and an intolerable disorder.

Chapter Eight: Treatment of TS

Because TS results from imbalances of neurotransmitters in the brain, the primary approach to controlling symptoms continues to be the administration of drugs that affect these chemicals. Various medications and techniques are used to treat the disorder, and although personal narrative appears in this discussion, in no way is the author either recommending or condemning the use of any specific treatment. Only a qualified physician can safely recommend methods for the management of Tourette syndrome. Today there are many new medications and behavioral therapies to help manage TS. Deep brain stimulation (DBS) devices are beginning to be implanted in order to treat severe cases of Tourette syndrome. These devices act in some way like a pacemaker for parts of the brain.

From 1961 until the mid 1980's, the drug of choice for the treatment of TS was often Haldol (haloperidol). Haldol belongs to the group of medications known as neuroleptics, antipsychotics, or major tranquilizers. It works by blocking dopamine receptors in the brain and is successful in about 80 percent of all cases in reducing tics. Many patients, upon beginning therapy with Haldol, notice an immediate improvement in their symptoms.

Although this drug is successful in diminishing tics, it is not well tolerated by some patients. Haldol is a powerful medication used primarily for the treatment of psychological disturbances. It, and even the other newer neuroleptics, can produce extremely unpleasant side effects such as cognitive blunting (decreased ability to learn), sedation, dysphoria (depression, irritability), weight gain, akathisia (restlessness), and nightmares.

Symptoms similar to those found in Parkinson's disease may also result from the long-term use of neuroleptics. These include muscle rigidity, tremors, and tardive dyskinesia, a potentially irreversible condition characterized by various unusual movements, particularly chewing motions. In order to minimize possible adverse effects while achieving the maximum benefits of neuroleptic drugs, the physician must carefully monitor the patient's response to treatment.

When I first went to a neurologist and received a diagnosis of Tourette syndrome, both the doctor's diagnostic manner and his way of prescribing medication struck me as peculiar. After diagnosing my problem, he brought an intern into the examining room. "My new intern has never seen anyone with Tourette syndrome," he said. "Could you do a tic for him?"

I tried to describe my tics to the young doctor but explained to both of them that I could not tic on command. It does not work that way. I felt that this was a very unusual request. Surely this experienced doctor knew that TS patients cannot voluntarily perform motor or vocal tics like a dog doing tricks.

After a brief reflex examination and a few more questions concerning my symptoms, I was prescribed Haldol in a rather loose fashion. I was told that I could alter the dosage to some degree, according to the severity of my symptoms. The doctor said that I might experience some side effects, and that I could lower the dosage if needed. It was basically a situation involving the lesser of two evils, the disorder versus the treatment.

After taking only a small dosage for a day or two, I became zombie-like, unable to think clearly. I could not perform any task that required concentration and I suffered from both fatigue and restlessness at the same time. Despite these problems, I continued with the medication for several months, determined to endure whatever was necessary in order to keep the tics at bay. The tics did diminish considerably, but only at the cost of my ability to function. I remember looking at the clock one day. It was two o'clock. I could clearly read the time but in my confused state I wondered, "What in the world does two o'clock mean?"

I eventually decided to try another neurologist. He confirmed the diagnosis of TS, but also preferred Haldol. He attempted to adjust my dosage, assuming that I might be taking too much. I tried the new dosage for almost two years but continued to feel like a mindless automaton. By this time I had given up any kind of work that required serious concentration. I was in a constant state of confusion, unable to perform the simplest of tasks. I was also experiencing tremendous depression and a tendency to nod off to sleep while driving. I realized that I had to change my medication. I had to find a doctor who would offer alternate therapies because Haldol was not the right drug for me.

Although my next physician agreed that neuroleptics were not a feasible treatment option in my case, the pointed out that these drugs are effective in the management of TS in many individuals. Other medications of this type include Orap (pimozide), Prolixin (fluphenazine), and Risperdal (risperidone). These compounds work in much the same way as Haldol but seem to produce fewer side effects in some patients. There are now much-improved neuroleptic drugs yet their safe and tolerable applications must be carefully determined by a doctor.

Another drug used to treat Tourette syndrome is clonidine. The brand name of clonidine is Catapres and it has been used as an antihypertensive (high blood pressure medication) for years. It has also proven to be effective in reducing TS symptoms in over half the patients who have tried it, particularly those with mild to moderate cases.

Symptom relief usually occurs more gradually than with neuroleptics, but the side effects are often much less bothersome. Clonidine's reported side effects include fatigue, racing thoughts, insomnia, low blood pressure, and headaches. Clonidine's role in treating TS centers around its effect on the neurotransmitter norepinephrine. The drug inhibits the production of norepinephrine in parts of the brain, which in turn alters serotonin and dopamine levels, and is usually administered in either pill form or as a skin patch. Many doctors prefer the skin patch because it allows the drug to be slowly absorbed, providing an even dosage throughout the day. Some patients must take the medication in pill form because of irritation from wearing the patch.

Due to potentially dangerous withdrawal symptoms, a patient cannot suddenly stop taking clonidine after prolonged use. Under the guidance of a physician, the dosage must slowly be reduced, usually over a period of weeks. Clonidine initially proved to be a godsend in my case. During my Haldol years, I had grown to believe that I was doomed to spend the rest of my life either as a drugged-out zombie or an untreatable sufferer of Tourette syndrome.

Finally, I was informed about an organization called The Tourette Syndrome Association, Inc. (TSA). I contacted them for help. They sent a list of pamphlets and medical publications containing information about TS. I was relieved to find that many drugs were being used for the management of this disorder, not just neuroleptics. After reading the literature I became interested in clonidine. The information included several interviews with patients who were taking the medication. Their accounts of clonidine's effectiveness in reducing tics were encouraging. Complaints of side effects were minimal compared to some other drugs.

I took the information to my family doctor, who was interested and supportive. He agreed that we should try clonidine. He carefully monitored my blood pressure as we gradually increased the dosage. Within a few weeks my tic symptoms began to diminish. The drug also had a calming effect, reducing my tendency to become frustrated or angry. For the first time in my life I felt significant relief from TS symptoms with no severe impairment of function. I took clonidine for about six years with considerable benefit and few side effects. The worst consequence was a feeling of fatigue, even when I was rested.

Eventually, I noticed a reduction in the drug's benefit but no decrease in the fatiguing side effects. My doctor suggested that I slowly reduce and finally discontinue the drug. My tics increased but my exhaustion evaporated. Only time will tell if I will continue to function well without the medication. Like virtually all drugs, clonidine is not perfect. All things considered, it was at that time one of the most well-tolerated and effective drugs used in the treatment of TS. Today, the choices are much greater, and the development of new medications has eliminated the use of many of the older drugs that I was prescribed as a young man.

There are also numerous antidepressants that are useful in controlling the non-tic symptoms associated with Tourette syndrome. Patients with obsessive-compulsive symptoms or depression may benefit tremendously from these medications. The use of antidepressants in the management of TS, OCD, and related disorders has increased dramatically in the past few years. Due to the various physiological effects of anti-depressants, extreme caution must be used if one type is discontinued and another started. For example, one family of antidepressants (MAO inhibitors) can clash with other drugs and certain foods, resulting in serious complications. It is important to discuss the use of any previous medication with a doctor before changing drugs.

One major group of antidepressants is referred to as serotonin re-uptake blockers, serotonin re-uptake inhibitors, or selective serotonin re-uptake inhibitors (SSRI's). These medications work by inhibiting the reabsorption of serotonin at the gap between neurons, the synapse. In addition to controlling depression, SSRI's are particularly beneficial in managing the symptoms of obsessive-compulsive disorder. Reported side effects include headache, lack of sexual desire, sexual dysfunction, insomnia, and loss of appetite.

Luvox (fluvoxamine), Zoloft (sertraline), Serzone (nefazodone), and Paxil (paroxetine) are among the many drugs in this class, but perhaps the best known of all serotonin re-uptake inhibitors is Prozac (fluoxetine). Every drug in the SSRI family is different, but each centers its effect on serotonin.

A relatively new drug, Effexor (venlafaxine), is both an SSRI and a norepinephrine re-uptake inhibitor. This drug is also proving to be useful in the treatment of depression and obsessive-compulsive disorder.

I have taken Prozac for several years and feel that it has been an effective tool in calming both my obsessive-compulsive and depressive symptoms. I also seem to cope with stress more effectively than I did before taking Prozac. This benefit can indirectly help control tics because TS symptoms are often amplified by stress. I have suffered no major side effects from this drug. My biggest complaint about Prozac is that it has caused somewhat of a reduction in creativity for me; it is harder to write songs since I have been on the drug. I have also heard others refer to it as a "motivation killer."

Another group of antidepressants, known as tricyclics, is also used to treat TS. One of these drugs, Anafranil (clomipramine), has been used in Europe for years to control obsessive-compulsive symptoms and is now being used in the U.S. as well. Tofranil (imipramine) and Norpramin (desipramine) are also commonly used tricyclics. These drugs are particularly effective against symptoms of TS associated with attention deficit hyperactivity disorder. Because they are antidepressants, they also aid in controlling the depression often seen in TS patients.

Drugs in other categories, such as Klonopin (clonazepam), Lithium (lithium carbonate),Valium (diazepam), and Inderal (propranalol) have also been used successfully in the management of TS. Most drugs used to treat TS were originally developed and approved to treat other disorders and later found to be effective against tics and obsessions. The list of treatments for Tourette syndrome is growing every year, as more existing drugs prove to be beneficial and new medications are discovered. Because individuals react so differently to drugs, months or years of trial and error may be necessary in order to find the best medication. No patient wants to be a guinea pig, but experimentation is sometimes unavoidable.

In the past few years, a debate has continued among physicians over the use of stimulants to treat children with attention deficit hyperactivity disorder. Stimulants can have a calming effect in pediatric patients but some researchers claim that they can trigger the onset of tics or even full-blown Tourette syndrome in certain people. Others argue that the hyperactive child who develops tics was destined to do so with or without the use of the stimulants.

Although ADHD is somewhat prevalent among TS patients, the reverse is not true. In hyperactive children who do not have TS, stimulant therapy can be beneficial. Caution should be used in treating any patient with ADHD who exhibits tics, as most researchers agree that stimulant use may cause preexisting tic symptoms to become worse.

Because TS can cause socially unacceptable behavior, a patient may develop problems coping with the resulting rejection and embarrassment. Feelings of low self-esteem, guilt, anger, and frustration often accompany a disorder, placing a person in an awkward position every day. The assumption that others are watching every move and making negative judgments, whether true or not, often creates emotional trauma. Psychotherapy can prove beneficial for patients experiencing these feelings.

Psychologists also offer methods for managing the primary symptoms of TS. These techniques (such as substituting an oncoming tic with a less obvious one) are extremely helpful in many cases. Breaking bad habits, reducing stress, challenging obsessive-compulsive urges, and developing relaxation techniques can all have therapeutic benefits. While psychotherapy alone is generally not successful in stopping tics, its use in conjunction with proper medication may prove to be an effective method of treatment.

Family members of the patient may also benefit from counseling. Feelings of guilt, shame, and anger commonly affect those who must deal with the frustrations of raising (or growing up with) a Tourettic child. A qualified therapist may be able to help parents better understand and cope with the inner mental struggles their child must face daily.

Openly discussing personal experiences with others affected by TS can also be helpful. The TSA and other organizations sponsor chapters in most major cities, many of which schedule regular support group meetings. People with Tourette syndrome frequently face their most difficult times as school children. Behavioral therapy combined with specialized educational techniques can be of considerable help in dealing with the social and learning difficulties experienced by many Tourettic students.

Although traditional Western treatments are the most widely used in the United States, there are alternatives available to the adventurous TS patient. Around the world, Chinese herbal medicine and homeopathic medicine are among the oldest and most widely practiced of the healing arts. Both of these disciplines operate on a different set of principles than those of conventional Western medicine. Concentration is focused much less on quick cures for specific symptoms and more on the patient's overall health. Emphasis is placed on strengthening and balancing the vital forces within, allowing the body to heal itself.

Chinese herbal medicine refers to this vital essence as qi (pronounced "chee"). According to ancient philosophy, qi permeates the universe and enters the body through the intake of food, liquids, and air, directly affecting the state of a person's health. By achieving an internal balance between opposing forces of nature, known as yin and yang, and by acquiring proper communication within the body among the natural elements of fire, wood, water, earth, and metal, the patient attains a strong qi, resulting in emotional and physical wellbeing.

When this balance is upset, disease occurs. The goal of the herbalist is to encourage the qi to flow freely by correcting the sources of imbalance. Proper balance is ideally achieved by lifestyle and diet. If needed, the herbalist may prepare a tea or potion composed of various herbs, minerals, barks, berries, or powdered animal parts to aid in the healing process. Interestingly, several of these substances have found their way into the pharmacopoeia of conventional Western medicine. Although the philosophy of qi is far from the mainstream in Western cultures, patients with a variety of ailments have claimed to find relief in the steeped concoctions prescribed by their venerable Chinese herbalists.

After years of conventional treatment, I decided to give Eastern medicine a try. I was told of an herbal medical center in a large city nearby, so I made an appointment. The walls of the small office were lined with hundreds of jars labeled with Chinese characters and filled with various substances ranging from whole strips of tree bark to insect parts.The herbalist, who spoke poor English, asked me to find the name of my disorder in a book and point it out to him. He then tapped his finger on my pulse, looked into my eyes, and examined my tongue. His diagnosis was that my earth and wood elements, representing specific organs and emotions, were not communicating properly. He prescribed a tea made from herbs, bark, and several indescribable components.

I drank the foul-tasting tea for a few weeks, noticing no relief from symptoms. At the time, my doctor had prescribed dosage changes in my regular medication as well, which may have diminished the herbal treatment's benefits. Someday I may attempt the herbal method again and stick with the regimen for several months. The original experiment was interesting but inconclusive. For now, I continue to be in the caring hands, or as some might say, the ferocious grip of traditional Western medicine.

Homeopathy is an alternative medical practice that has also been used in the treatment of Tourette's. Homeopathic medicines are sold in accordance with a separate pharmacopoeia, which is FDA approved. Proponents of this type of treatment say that the medicines evoke a curative response that allows the body to gradually heal the condition.

Alternative disciplines for the management of TS also include nutrient therapies and the avoidance of allergens. Nutrients recommended for TS patients are believed to improve overall brain function, increase oxygenation of brain tissues, and help rid the brain cells and bloodstream of lipopfuscin, an unwanted byproduct of fats and unattached oxygen molecules, known as "free radicals." Recommended nutrients are also believed to affect the balance of neurotransmitters within the brain.

Researchers have found that allergic reactions to certain foods and irritants may also increase tics. Alteration of diet, lifestyle changes, and a regimen of fasting may lessen these negative responses.

Many forms of alternative medicine exist and some may ultimately prove superior to conventional treatments. Unfortunately, a current lack of scientific data prevents researchers from adequately documenting the effectiveness of alternative therapies. Perhaps in the future this informational void will be filled.

Deep Brain Stimulation (DBS) is a surgical treatment in which a device is implanted to help regulate certain brain functions. Patients with severe tics have found incredible relief from this implanted device. However, this is a relatively new and very expensive approach. Its long-term benefits are still being evaluated. Hopefully, it will become available and affordable to more patients.

Chapter Nine: Dealing With Society

I was sitting in a restaurant trying to eat a sandwich when a young man approached me and began rapidly asking questions: "Hey man, what are you on? What drug are you coming down from? Do you need help? Are you sure you're okay?" I thanked him for his concern and assured him that I was not suffering from drug withdrawal and did not need help. He assumed, as have many others, that a person jerking about and shaking must be having a reaction to a street drug like cocaine or heroin.

My odd table manners often seem to fuel this suspicion. Due to tics, I constantly spill things or shake the food off my fork. In order to avoid embarrassment when eating out, I frequently try to find a hiding place such as a seat located in the rear of the restaurant. It is difficult enough to control tics while eating at home; I do not need the added stress of an audience.

Sometimes circumstances make it impossible to escape from people, their curious stares and inappropriate reactions. A few years ago, I was singing a vocal track in a recording studio. Much to my surprise and embarrassment, the engineer recorded my grunting between verses then played it back to the musicians in the control room. I had been wearing headphones during the recording process and I had no idea that I was having vocal tics. Everyone in the control room thought that this joke was funny but I found it to be tasteless and irritating. It was difficult for me to control my temper, but I finally managed to laugh it off.

Public displays of Tourettic behavior can sometimes lead to more serious problems. In my college years, I was quietly walking across a campus parking lot on my way to class. The police pulled up and said they wanted to speak with me. They asked several questions about what I was doing, where I had been, and where I was going. The officers commented that I appeared to be extremely nervous and they questioned the reasons for my jerky movements. At the time I did not yet know the cause of my fidgeting and I was unable to give them a satisfactory answer. After a few minutes they evidently decided that I was harmless, thanked me for my cooperation, and moved on.

Understandably, it was upsetting that the police singled me out of thousands of college students for questioning. I assume that they saw me ticcing and thought that I might be either dangerously crazy or on drugs. Fortunately, the only consequence of this particular encounter was that I was late for class. I have heard of others with TS who were actually held in custody by the police until the misunderstanding was cleared up.

People generally feel either aversion or sympathy for a person with tics. The sympathetic onlookers are genuine in their concern, and I appreciate their interest in my wellbeing. Nevertheless, the constant approach and inquiry of strangers can be irritating. In the past, I would become depressed and withdrawn whenever my odd behavior was questioned. From time to time I would stay at home, feeling too embarrassed to go out in public. A simple task, such as a trip to the grocery store, was a dreaded experience. When the tics were unmanageable, they would draw attention like a magnet.

I have since gotten over some of the fear of ticcing in public yet I continue to be upset at times by the reactions of others. Naturally, I would prefer that my tics not be noticed and I continue to disguise them whenever possible. Like an "undercover Touretter," an alien among normal citizens, I try not to tic and blow my cover.

There are those less fortunate who are unable to hide their tics at all. They cannot go undercover and must endure the scorn of others daily. The constant bombardment of disapproving looks, endless questions, and outright harassment eventually becomes unbearable. Sometimes, withdrawal from society seems to be the only viable course of action.

The early years are often the hardest as Tourettic children sometimes suffer mercilessly at the hands of their peers, who have no understanding of complex neurological disorders. Children simply brand a person who acts differently as an oddball, a good target for teasing. Numerous horror stories are told of children with tics who are relentlessly tormented by their classmates, making life at school a living hell.

For the young, retaliation is only natural. It is not unusual for Tourettic students to end up in trouble for physically attacking their tormentors. This reaction may be extreme, but the anger is understandable. I often feel the urge to tell judgmental adults to "go to hell." Instead, I usually explain my disorder, giving them an opportunity to learn something about Tourette syndrome. It has been my experience that people who were initially critical may actually appreciate the explanation and grow to accept my tics.

I am fortunate that in my profession, many people seem to accept me the way I am. Being a musician is almost like having a license to act strangely, as artists and musicians have been social misfits throughout history. Wolfgang Amadeus Mozart is believed to have had TS, and Vincent van Gogh was epileptic. Most people seem to embrace such odd behavior as a part of being creative, a characteristic of the quintessential artist.

It is unfortunate that this acceptance does not cross into the business world so easily. Business people frequently have a rigid concept of professional poise, and a person with tics does not fit this image. In the corporate world, being respected is of far greater benefit than being liked. Most professions involve some degree of presentation or show, and the performance required by those in other fields is usually more demanding than that of a musician. A corporate executive must appear totally composed at all times in order to avoid suspicions of incompetence. Politicians must seem to have virtually no flaws as humans or leaders.

Due to a fear of losing this essential respect, I avoid business meetings whenever possible, preferring to let someone else handle negotiations. Executives in the recording industry expect an artist to be totally confident and calm, not fidgety and withdrawn. Once a performer becomes famous, the artist's strange behavior is miraculously tolerated, perhaps even viewed as cool or expressive. But until then, the unusual artist is often rejected by business people before they even know anything about the person.

In retrospect, one of my most awkward tic disasters is now something I can laugh about. I was in a band that ended up signed to a major record label in the mid 1980's. As all major record company folks did in those days, the label people wined and dined us (with our own royalty money) in an effort to show us how cool and important they were. We were taken to a top-level restaurant and someone ordered flaming drinks for all of us. About the time the waiter placed my glass on the table, I had a major tic with my arm. This sent burning alcohol in a tsunami of flames rolling across the table. Not only did everyone at our table scramble, half of the diners jumped up in horror as the waiters rushed to put out the fire. Of course, I was embarrassed to the core. However, after looking back on the lame treatment we received from this record label, I am now glad it happened. The record company executives did not look so cool and important while running in a panic from a sea of flaming liquid.

For those with tics, the inability to exhibit this poise can present a severe handicap, both socially and financially. Although some individuals with TS manage to excel in mainstream professions, most suffer discrimination. Many lose their jobs due to Tourettic behavior. Others are not considered for well-deserved promotions as a result of their humiliating tics.

It is fortunate that many people with TS develop their own effective methods of coping with the disorder. The majority are highly driven individuals, perhaps out of necessity. Still, a severe case can overpower even the most motivated person, transforming life into a series of miserable, explosive tics. There are many diseases that ravage the body much more severely than Tourette syndrome, but few attack the soul with such vengeance. A patient may be forced to resort to extreme means such as constant sedation by powerful drugs, or total isolation. In a few cases, distraught TS patients have undergone psychosurgery in a desperate attempt to rid themselves of tics and obsessions. These older surgical procedures are controversial, experimental, and at best, extremely risky. Fortunately now, there is a relatively new procedure called deep brain stimulation that is showing great promise.

For those with moderate cases, tolerating the disorder is usually less of a problem than dealing with the people who make no attempt to understand it. I am thankful that I have been surrounded by people who do not think negatively of me because of my symptoms. I would much rather have the tics of Tourette syndrome and not be judged for them than be totally healthy yet besieged by judgmental people who criticize my every action. Most people would probably feel the same way under similar circumstances.

Imagine attending a quiet lecture while suffering a bad cold, sniffing at least ten times a minute, unable to stop. No medication will help the symptoms and missing this important event is not an option. What would bother you the most - the fact that you are sniffing, or the thought that everyone else may be annoyed by your sniffing? I would bet on the latter. Similar uneasiness is experienced by a person with TS every day, every hour, and in every public location.

Relatives, co-workers, and peers sometimes either lack the intelligence or the humanity to accept tics; many cannot, or will not, understand that inappropriate behavior can be caused by an organic disorder. They assume that if the patient can control tics for short periods of time, he or she should be able to keep them permanently at bay, or at least do so whenever in public. It may appear that the person's tics indicate a lack of both self-control and willpower.

It is often forgotten that tics are just as necessary to those with TS as the act of sneezing is to victims of allergy attacks. Most people can hold off a sneeze for a few seconds, maybe longer, but inevitably an uncomfortable sensation will force them to sneeze. The act of giving in to a sneeze does not suggest that the allergy sufferer has no willpower. The same may be said for anyone trying to restrain from vomiting, coughing, or any number of bodily functions. Absolute control is impossible. In the case of tics, the speed and unpredictability with which they occur makes suppression even more difficult.

The Tourettic mind is believed to contain our most primitive animalistic tendencies, emerging in an uninhibited, twisted form. Like a jungle cat quietly waiting to pounce on its prey, the disorder may unexpectedly roar into action at any time, turning a brief moment of tranquility into a nightmare. The psyche of Tourette syndrome possesses at least as much intelligence as its host and it is significantly more cunning. This combination of the primitive and the intelligent causes sudden, explosive, yet complex and sometimes uncouth tics.

Other chemical disturbances, such as intoxication by street drugs or alcohol, can also cause both primitive and uninhibited reactions. The symptoms of TS are different from those associated with drunkenness, but the source of the behavioral alteration remains nevertheless chemical. The behaviors induced by TS and other disorders, however, do not require the introduction of outside toxins. The public seems to understand and accept that reprehensible behavior can result from alcohol and drug use, but people have difficulty comprehending that it can be caused by built-in imbalances. The self-indulgent drunk is readily forgiven yet the blameless TS sufferer is shunned.

This lack of understanding applies to many biological conditions. Consider the diabetic who begins to act irrationally during the early stages of diabetic coma. Due to a chemical imbalance, this person temporarily loses control over the mind and body. There are hundreds of documented cases in which people with diabetes were denied medical attention because others assumed that they were merely drunk from alcohol. Many of these unfortunate individuals died in hospitals as they awaited emergency treatment.

Just as an imbalance of insulin occurs with diabetes, an imbalance of neurotransmitters causes Tourette syndrome. The two are completely different physical problems yet both are chemical in origin and both can produce bizarre symptoms that falsely appear to be either psychologically based or caused by drug abuse. In either case, misinterpretation of outward symptoms can have destructive effects.

Most people with Tourette syndrome do not expect, nor should they, any special treatment. What they should and do expect is to be treated with dignity and equality. Basic human understanding can do wonders to help those with TS, and the lack of it can be devastating, especially to children. Many develop life-long psychological problems as a result of mistreatment by the people around them.

Perhaps someday members of the public will learn to accept the tics of Tourette syndrome as readily as they do the sneeze of a common cold. Until then, the old cliché "what you see is what you get" may be the best mode of social interaction for those with TS. With or without tics, a person is either accepted or rejected by others, and one must be able to face both possibilities.

Chapter Ten: Final Thoughts on The Unwelcome Companion

One afternoon I was having a particularly difficult time with tics while setting up my musical equipment for a performance. A young man approached me with an astonished expression. "My little cousin does that," he said. "He's been kicked out of school and his parents have practically disowned him. I'm the only one who will have anything to do with him."

I was struck by feelings of pity for this child and could not help but feel anger toward his parents and school officials, in spite of their ignorance. He described his cousin's tics, and I explained that the symptoms sounded like the child could have Tourette syndrome. I suggested that he immediately take the boy to see a doctor for a proper diagnosis. I also gave him my phone number and offered to provide the address of the Tourette Syndrome Association as well as any other information I could find. I never heard from him again. I hope things worked out well and this child's disorder is now being properly treated.

While traveling in Italy, I saw a middle-aged man displaying severe motor and vocal tics. He was receiving strange looks from nearby pedestrians as he attempted to gain control and continue moving down the sidewalk. He would take several steps, yelp, and then become frozen in contortions. His arm would flail and his torso would twist. I was upset by this sight and wanted to communicate with him but I could not speak the language. I also feared that my attempt would cause him additional embarrassment. It was difficult to walk away and do nothing because I suspected that, like so many others, he may not have had a clue as to the nature of his problem.

Because I am not a doctor or scientist, I feel somewhat powerless to help those who are devastated by this disorder. Perhaps this book will make a small contribution in the battle against TS by waving another distress flag at those who can provide direct assistance.

As TS research continues, scientists will inevitably create new and better methods for treating the disorder, improving the odds for abatement of symptoms or recovery in even the most severe cases. Although Tourette syndrome's possible genetic connections to other disorders make it extremely perplexing, these links may eventually prove to be beneficial. Every study of a possibly related condition may provide clues to unravel the TS enigma.

While researchers do biological battle with Tourette syndrome, the average person can help the cause by educating others. One source of the disorder's power is its ability to humiliate and embarrass its victims to the point of ruining their social and professional lives. If the public could become accepting enough to make such humiliation obsolete, a great deal of the disorder's strength would disappear. TS will likely always be a disruptive neurological problem but it need not be a socioeconomic handicap. In the real world, people tend to discriminate against any variance from the norm, particularly when they don't understand it or perceive it as threatening. Until this situation improves or a cure is discovered, those with TS must learn to either tolerate or ignore the judgment of others.

The extensive list of symptoms associated with Tourette syndrome, both direct and peripheral, is continuing to expand, largely because of the growing number of cases studied. Many of these symptoms are variations on similar themes, such as the multitude of obsessive-compulsive behaviors. Others are new and puzzling discoveries. It is easy to become overwhelmed by the number of possible problems that may either result from or tend to co-exist with this disorder,. Fortunately, patients rarely exhibit all of the known symptoms and many suffer only a few.

Although any disease or disorder is an enemy to its victim, Tourette syndrome excels in its ability to whittle away at the sufferer's self image. In contrast to many other behavior-altering disorders, TS does not cause a person to lose consciousness or awareness of outside reality. One can never retreat to or become content in a private little world. It is a never-ending struggle to avoid total chaos as the Tourette world clashes with the real one. The battle is exhausting and the casualties are many, including losses in self-esteem, productivity, confidence, and personal relationships.

There is a measure of hope for many children with tics, with or without treatment. It is estimated that in approximately one-third of all tic disorder patients, the teenage years bring a remission of symptoms. The tics and obsessions may never return. In about half of all other cases, they may not disappear but the symptoms will significantly decrease by late adolescence.

The remaining patients will probably have symptoms of relatively constant severity throughout life. Even in these cases the disorder is unpredictable, and the symptoms may spontaneously improve at any time. TS usually reaches its peak of severity during early adulthood and rarely becomes significantly worse with age. It is far more likely that tics will fade with time.

All of my adult life I have witnessed people who seem to be able to relax and allow their stress to disappear for a while. They can quietly enjoy a book, drink a cup of coffee with a friend, or watch a movie while showing no sign of nervous tension. This is just a pleasant part of life that most humans take for granted. Such simple repose is both foreign and truly amazing to me and I assume that it is to others who have TS. As much as we would love to be able to relax, we may never actually realize such an ability.

The only opportunity we have to rest must sometimes be purchased in the form of a sedative medication. Without pharmaceuticals, we are either focused and working, ticcing, or both. With the help of proper drugs we may reach a state of synthetic relaxation, but only at the price of being troubled by the medication's unwanted side effects.

To most people, the idea of simple relaxation as a major achievement seems silly. Yet for those of us who live with the onslaught of tics, swimming the English Channel is probably a more realistic goal than sitting calmly on a sofa for half an hour.

Even when we are trying to relax, the conflict rages inside our minds, demanding enormous amounts of energy. The rapid-fire of intrusive thoughts, the blitz of urges and obsessions, the tensing of muscles as they launch into frenzies of tics - all of this is exhausting. A person with TS may appear to be sitting calmly, revealing only an occasional twitch or blink; but behind those eyes roars a siege, an electrical storm, a violent confrontation between the understandable and the unthinkable - and so it goes forever.

A person can become so obsessed with Tourette syndrome that it consumes one's mind. A developing cycle of obsessions, compulsions, tics, embarrassment, rejection, shame, frustration, and depression can create a whirlpool of despair. Sometimes the only way out of this vortex is simply to ignore the "demon" whenever possible. To ignore it is to rob it of its power.

I cannot stop the tics and compulsions. They happen. I perform them; but I try to quickly move on. Goals and dreams come first. I cannot allow the Tourettic demon to dominate my thoughts. Although the disorder can cause enormous harm, positive and productive characteristics may also result from it. People with TS often excel in creativity, intellect, sports, and musical ability. Although some patients display inferior handwriting skills and other fine motor coordination problems, many exhibit superb reflexes, speed of movement, and agility.

Baseball legend and Tourette patient Jim Eisenreich has become a major spokesperson in the campaign for awareness. His struggle with the disorder and his athletic achievements serve as an inspiration to us all. Mahmoud Abdul Rauf is an All-American basketball player with TS. Tics did not stop him from setting several NCAA records and becoming a professional ballplayer. Others throughout history have achieved incredible feats in spite of their obvious tics. Investigators have turned up evidence that Peter the Great and Napoleon Bonaparte, two of history's greatest military leaders, probably suffered from TS. They each rose to power in the days when people with neurological disorders were often locked away or burned at the stake as witches.

Friends and relatives of the great composer Mozart described his eccentric behavior in detail. His symptoms apparently involved a vast array of complex motor and vocal tics, including coprographia and coprolalia. He was also well known for his elaborate obsessive-compulsive rituals.

Dr. Samuel Johnson, noted intellect of the 1700's, apparently had a severe case of Tourette syndrome. Johnson is best remembered as a lexicographer and author of "A Dictionary of the English

Language," published in 1755. Several eyewitness accounts recorded by his associates describe the complex details of Dr. Johnson's bizarre ritualistic actions. There was much speculation as to the cause; Tourette syndrome was yet to be identified. In spite of this, Dr. Johnson was greatly admired and became known as "the greatest man of his time."

The same villain that creates internal conflict can provide strength in the external struggle for survival. The world outside may be easier to conquer than the realm of one's own mind. True inner peace is something that many people with Tourette syndrome may never know. Most of us would give everything we own just to be able to relax, sleep, work, and play without the constant torture of this disorder. We learn to savor every moment of tranquility that life has to offer. During those rare moments, when the demon sleeps, we get a brief but precious glimpse of serenity.

Part Two:

Beyond The Unwelcome Companion

Chapter Eleven: The State of Tourette Syndrome, Then and Now

When writing my first book on the subject, The Unwelcome Companion, An Insider's View of Tourette Syndrome, I felt that a basic layperson's guide to the disorder was needed. At the time, many excellent books by both physicians and patients were available but the area of "how it feels to experience Tourette syndrome" had not been adequately covered. The book was an attempt to help address that problem. The Unwelcome Companion in its original form continues to be well received and I am most appreciative to all who have read it.

I realize that I cannot possibly write, nor should I try to compile, a comprehensive book on the medical treatments and the biological components involved with this disorder. Even if I were qualified to do so, such a text would require constant revision in order to keep up with the latest research. For those who would like to learn more about the science behind the symptoms and treatment of Tourette syndrome, there are many excellent sources written by professionals in the medical and research fields currently available.

It is up to those of us who actually have Tourette syndrome to attempt to reveal its inner secrets, including the mysterious origin and volition of tics, the crippling impact of obsessive-compulsive rituals, and the hidden sensations experienced with this disorder. Of course, almost every disorder affects every individual somewhat differently. I can only speak for myself and I would never profess that my experiences are absolutely definitive of anyone else's. My discussions with other people who have TS have lead me to believe that many of us do indeed experience very similar mental and physical events.

It is interesting to discuss TS with others who have the disorder. Every person has developed his or her own coping mechanisms. I have met people who actually embrace their Tourettic symptoms and their whole persona becomes one with the disorder. Some even claim that Tourette syndrome is a blessing.

If such views help these individuals to cope with their tics, I am happy for them. Furthermore, perhaps they are absolutely correct in dealing with the disorder in such a manner. For some, TS may indeed be a blessing. Whether their attitudes are borne of delusion or brilliance is not for me to say.

I consider Tourette syndrome to be a negative factor in my life. It has cost my family and I thousands of dollars in medical bills over the years and it has repeatedly helped to damage me professionally.

In addition to the primary degradations of the disorder, my tics have also resulted a plethora of self-injuries and other physical problems. At this moment, I can feel the pain of a jaw that closes too far because of my tooth-grinding and chomping tics. My dentist has informed me that I will soon need to have more major dental work done in order to repair this problem.

In spite of my negative feelings about having TS, I honestly do not have a reason to complain. I live a good life in spite of the problems caused by Tourette syndrome. I have a moderate case of the disorder and I am lucky enough to have a great support system of loving people. I am also aware that people face much worse problems on a regular basis, including life-threatening diseases. I am also thankfully able to prevent Tourette syndrome from defining me. If it can, it will define a person.

Unfortunately many people with TS are not so lucky. Some people have such severe Tourettic symptoms that they are completely debilitated. Many must face the wrath of this disorder without the benefit of supportive families and friends.

Although a wealth of information is emerging about the mechanics of the brain and new treatments for TS are sprouting from this knowledge every year, it seems that even some of the brightest minds are totally missing the mark when it comes to truly understanding Tourette syndrome.

Until those of us who experience TS on a daily basis start telling all, the chemists, counselors, and physicians will have to search much harder to find help for us. Unthinkable intrusive thoughts, debilitating frustration, self-hatred, and explosive internal rage are among the more disturbing problems that many of us with this disorder face. These conditions are much less addressed by caregivers primarily because they are not often discussed by people with TS. It is also sometimes unclear if these symptoms are the product of another co-existing disorder or if they actually result solely from Tourette syndrome.

Many times the person with TS is either too confused or too embarrassed to reveal the most troubling manifestations of the disorder. There is also a tragic tendency among some who do not have TS to whitewash Tourette syndrome in general. Some try to paint it as a humorous and harmless little malady. They somehow cannot face the possibility that their friend or loved one is really having a hard time because of Tourette syndrome. This denial makes the person who does happen to suffer disturbing symptoms feel even more awkward about bringing them up for discussion.

TS is certainly not the end of the world for anyone, and most people with TS live happy, productive lives regardless of their tics. However, both disregarding the disorder's potential and discouraging the discussion of its deepest disturbances only increases its power to do harm.

In the remaining chapters, I will attempt to describe some of these less discussed, often distressing symptoms that seem to occur regularly in many people with TS. Some readers may find parts of the text quite dark and somewhat frightening. However, this book reflects the truth as I see it personally. The descriptions herein either came from self-observation or from the stories shared with me by others who have TS. My intention is to promote a better understanding of this disorder, and I encourage others with Tourette syndrome to speak out about their experiences.

The reader should keep in mind that people with Tourette syndrome are incredibly strong. We grow up fighting this monster every minute of every day and we typically handle things quite well. I know many other people with TS and virtually all of them live very good lives.

Chapter Twelve: The Anti-self

Although Tourette syndrome is not a life-threatening disorder, I feel that my Tourettic mind, my "anti-self," would totally destroy me if it could. Fortunately I can keep it from doing so, but those with severe cases of TS often suffer substantial physical and mental damage from the relentless symptoms.

A person with TS will soon discover that the anti-self is only concerned with destruction. It has absolutely no good intentions; it is ruthless. It clearly foresees the negative consequences of its actions yet it attempts to carry them out anyway. It is fully aware that its aggressions may cause catastrophic results for the affected person yet it is incapable of feeling any guilt whatsoever.

Caring for anyone or anything is not within its comprehension and any collateral damage is of no concern to the anti-self. It harbors no regrets, even in the worst of outcomes, and it thoroughly enjoys its abilities to cause despair. If fact, the Tourettic mind is totally void of self-generated compassion. Much like the cruelest rulers of the Roman Empire, it believes that any merciful thought is both a sign of vulnerability and a detestable human weakness.

The anti-self can detect the presence of emotions felt by its host but it cannot feel them as emotions at all. They are merely perceived as markers indicating that the host may have yet another Achilles' heel.

The anti-self has no sincere feelings other than hostility toward its host, an unquenchable thirst for sabotage, and a sense of pleasure derived from committing sabotage. Virtually all of the sensations felt by the host are divided into either opportunities or non-opportunities for attack. The Tourettic mind is always searching for weaknesses in the host's defenses. At times it may feel wickedly playful and actually amuse the host for a short while. However this game can quickly turn ugly without warning.

It is odd that, in spite of tolerating this internal torment, most people with Tourette syndrome are extremely compassionate and understanding toward others. Perhaps this is because those of us with TS know how it feels to be bedeviled and never wish to do so to anyone else. It could be because we spend so much of our lives trying to understand ourselves that we dare not pass premature judgment on others. It may even be because we personally know and detest that which is hurtful and we fight it everyday, from the inside as well as from the outside. The Tourettic part of the mind is quite sinister but fortunately its attacks are usually only directed against the person in which it resides.

I have met several other individuals with TS and I feel that the last thing any of us want to do is to cause harm to anyone. We are generally sick and tired of the venom that lives inside us and toys with our every thought. We don't want to spread the devastation, we just want to try to ignore it and live peacefully.

People with TS in centuries past were almost never permitted to live in peace. They were easy targets for fanatical witch hunters, intolerant religious leaders, and perpetrators of superstition and folklore. In some cases, people with obvious symptoms were not allowed to live at all. Many people with TS are believed to have died either in asylums or at the hands of religious fanatics.

Although knowledge of neurological disorders was nonexistent in the year 1486, some researchers believe that the first written description of Tourettic symptoms appeared that year. During the infamous Holy Inquisition, a user's manual for identifying and prosecuting witches was written. It also included instructions for evaluating the demonically possessed. Entitled The Malleus Maleficarum (Der Hexenhammer or The Hammer of Witches), the book provided guidance for the Catholic Church and its supporters in the prosecution, torture, and sometimes even murder of thousands of innocent people.

Two Dominican monks, Jakob Sprenger and Heinrich Kramer, wrote the book at the request of Pope Innocent VIII. In one section, the authors described the symptoms of a priest who was possessed by the devil. It is now believed that the priest may have actually suffered from Tourette syndrome; but of course, there is no way of knowing. Unfortunately he too became convinced of his demonic possession at the time.

Translated from the Malleus Maleficarum:

"But when he passed any church, and genuflected in honour of the Glorious Virgin, the devil made him thrust his tongue far out of his mouth; and when he was asked whether he could not restrain himself from doing this, he answered: "I cannot help myself at all, for so he uses all my limbs and organs, my neck, my tongue, and my lungs, whenever he pleases, causing me to speak or to cry out; and I hear the words as if they were spoken by myself, but I am altogether unable to restrain them; and when I try to engage in prayer he attacks me more violently, thrusting out my tongue."

End of excerpt

It seems strange in these modern times to even consider a diagnosis of possession by evil spirits. I find the whole demonic idea laughable today but nevertheless I feel that these past conclusions were understandable, given the state of science and the overwhelming influence of religious dogma during the middle ages. I also believe that, to the so-called "possessed," these sensations felt very real.

In my opinion, a description of how it feels to have Tourette syndrome is not complete without emphasizing the sensation of being under the influence of an invading entity. It may start as a very vague sense. Eventually, however, a person with TS may feel as if he or she is fighting for control against something that is intelligent yet alien to his or her own consciousness. From my experiences, and from the discussions I have had with other TS patients, this feeling of fighting an entity can be a good way to explain how the disorder feels.

If I had lived in another era, I might have bought into the Church's "demoniac" idea and considered myself a victim of evil spirits because that is exactly what it often feels like. Why does it feel like possession? Primarily because the volition of tics indicates an unwanted influence.

In my opinion and in the opinion of many researchers, tics are not truly involuntary. This notion is contrary to what most people believe because they do not understand how and why someone would voluntarily tic. Tics are absolutely unwanted to the person's conscious mind; it does not volunteer to carry them out. They are, nevertheless, purposeful actions. These actions are solely created and performed by the Tourettic mind against the will of the conscious mind. This causes a constant struggle between the forces that want to perform the tic and the ones that resist performing it.

The conflict literally wears a person out, physically and mentally. People with TS commonly feel drained, exhausted from the battle.

Not only does the Tourettic mind order the performance of tics, it keeps tabs on whether the afflicted person performs them properly. If the host is able to control the tics for a while, the anti-self stores this information in memory and recalls it. It then attacks with even more ferocity, pressuring the host to perform an often uncontrollable, violent explosion of "past due" withheld tics.

Researchers have debated the voluntary versus involuntary aspects of tics for years. Perhaps many simple tics are indeed involuntary. On the other hand, maybe these tics are actually voluntary but the process happens so quickly that the premonitory urge is not evident to the affected person. It is very hard to tell.

Although it is difficult to always determine the true volition of tics, I feel that one can best describe most tics as "voluntary" but "undesired." This is why many of us with TS can temporarily control some tics, but not for long. Many of the simple eye blinks, grunts, and shrugs seem to just pop up and be quickly gone; but the more complex movements can be withheld by the person in a manner much like holding off an oncoming sneeze or cough.

As odd as it sounds, a tic cannot accurately be reproduced at will. This suggests that the tic is created wholly by the Tourettic mind.

For example, I cannot precisely imitate myself ticcing. It is impossible for me to deliberately and consciously execute a move without the motor action being a bit more fluid than an actual tic would be.

Sure, one can mimic the tic's movement or vocalization. However, upon close inspection, it is never exactly the same. It seems that the speed, force, and rhythm of many tics are calculated and executed using a formula that is alien to our conscious knowledge. In addition, the Tourettic tic seems to involve a slightly different set of muscle memories than the purposeful movement. In other words, the anti-self may produce somewhat different mannerisms and perhaps even different vocal sounds than those created by the conscious mind. These common odd mannerisms and sonic differences likely contributed to the misunderstanding of TS in the past.

The folly of our ancestors can teach us much about Gilles de la Tourette syndrome. Many of the symptoms described by "demonologists" of the past now almost perfectly match those of TS. Probably the most famous one is coprolalia (the uncontrollable uttering and sometimes shouting of profanity or socially unacceptable words and phrases). This symptom is particularly hard to understand even in modern times unless one considers the fact that it is unwanted yet somehow voluntary.

It is reasonable to ask, "Why would a person uncontrollably shout profanity or make socially taboo statements?" It is hard enough to comprehend the act of blurting out something unintentionally to begin with but it is even more difficult to fathom that the statement would have to be of a profane nature. Why would the Tourettic person not simply yell, "hello, dog, house, sandwich," or anything but profanity? Sometimes they do; but in order for the tic to be considered true coprolalia, the utterance must be of a profane or offensive nature, as indicated by the Greek term "coprolalia," meaning "feces language" (kopros = feces, lalia = to talk).

The answer to the mystery of why one uncontrollably yells profanity is at least somewhat understandable to scientists who study the inner workings of the human mind. Certain brain chemicals called "neurotransmitters" can either inhibit behavior or reduce the ability to inhibit behavior. The influence of these neurochemicals becomes more apparent when their balance is distorted due to toxins, drugs, or physical disorders such as Tourette syndrome.

Serotonin, for example, is a chemical known to aid with inhibition, while an overload of a different neurotransmitter, dopamine, can contribute to uninhibited behavior. A person whose brain does not utilize these chemicals properly may say the very thing that they know should be suppressed.

After the Tourettic mind goes through life's experiences with its host and observes all of its surroundings, it quickly learns wrong from right. It soon can distinguish between an appropriate act and a non-appropriate one. It has had years to develop its skill as a saboteur and it knows that vulgar or blatantly anti-social language will embarrass the host to a great degree. In fact, without a clear knowledge of what constitutes a social offense, the anti-self would be totally unable to perform the act of coprolalia. If a person with coprolalia grew up thinking that the word "tree" was vulgar, then he or she might inappropriately yell "tree," and this would indeed be an act of coprolalia, at least to that person.

Therapists once mistakenly assumed that coprolalia must always involve sexual connotations. It does not. It involves saying what is known to be detestable. The profane statement may turn out to be a sexual term, a racial slur, a religious insult, or anything else that is offensive.

There are other times when such a vocal outburst explodes randomly and is not triggered by anything in particular. Some words and phrases simply become habitual and are shouted whenever tension builds. Therefore the same word or phrase (obscene or not) may pop up over and over in a variety of settings. It may even become a hallmark of the person.

Although I am lucky that I do not have coprolalia, I fully understand where it is coming from. Most estimates say that less than one third of Tourettic people exhibit coprolalia yet the mechanisms of this odd symptom are much like those of other tics.

I may never uncontrollably utter a profane word, but I will inevitably have unwanted thoughts that are the opposite of what I know to be morally right, just, and especially socially correct. All too often these thoughts can cause enough tension to trigger physical tics.

I fully know that it is not acceptable to slap my hip repeatedly in public but sometimes I cannot stop it. I know that it is not proper to rapidly scrape the plate with my fork while attempting to eat yet it happens. This is much the same as coprolalia in that it is improper and repulsive. My conscious mind knows that the act is both nonsensical and socially suicidal. It is a very bad thing to do at the time; and because I know this, the Tourettic mind knows it as well.

Imagine trying to drive a defective automobile with two steering wheels. Every time you want to turn your wheel to the left, the other steering wheel guides the car to the right. You have to quickly reach over, grab the second wheel, and jerk it to the left as well in order to execute the left turn. In this situation, the second wheel and its opposite action would represent the Tourettic mind and its desire for sabotage. The act of reaching over and jerking the second steering wheel would be similar to the performance of a tic.

The above analogy applies even to a very quick, simple tic. Coprolalia and other more complex tics involve a similar scenario with the added knowledge that there will be significant consequences for committing the act. Significant consequences or embarrassment suffered by the host are exactly what the anti-self wants.

In the case of the unfortunate priest discussed in The Malleus Maleficarum, his symptoms became worse when nearing a church because his Tourettic mind knew that the church was the most inappropriate place for a person to act in a vulgar manner, particularly a person considered to be a man of God. The church was the absolute worst place for a priest to stick out his tongue, curse, and contort his body into strange positions. The Touretttic part of the priest's mind was gladly aware of this situation; that is exactly why it attacked the man when he was near a church. This 1486 account of the priest's tics can now be seen as a perfect example of Tourettic behavior.

This awareness by the ant-self also explains why some so-called possessed victims become more agitated in the presence of religious symbols or clergy. To most believers, their religion represents good, proper, and orderly behavior. The Tourettic mind is well aware of this and enjoys working extra hard to disrupt the very moment when the host should be behaving in a reverent manner.

"The idle mind is the devil's workshop," an accurate statement on many fronts, also applies to Tourette syndrome. While one is occupied with an activity, the tics diminish. When a person is idle, the symptoms come on full-force.

I have significantly less tics when I become engrossed in an activity such as playing a musical instrument. The many athletes with TS also say that vigorous mental and physical activity as well as focus can help them temporarily ward off their symptoms.

I try to benefit from this aspect of the disorder. I stay busy in order to avoid facing an onslaught of tics. If my mind is not focused, the obsessions and tics will move in immediately. Therefore I am at least always trying to be productive. However, I must admit that this constant activity is performed out of necessity and it is not an indication of any hard-working characteristics on my part.

This "idle mind" phenomenon has always been a problem for Tourette patients when they visit the doctor. During a very focused examination or an engaging conversation with the doctor, the person may not tic at all. They are tuned in and intensely discussing a serious matter. Therefore the Tourettic mind gets temporarily pushed back. It cannot exert enough power to push through.

The physician can erroneously assume that the patient has a much milder case of the disorder than he or she actually has. Sometimes this leads to a lack of treatment. If the patient is currently under treatment, the absence of tics may also lead to the doctor to mistakenly think that an existing medication is more effective against tic symptoms than it really is. This is an understandable situation and one that is common.

However, shockingly, there are still some horrible situations that are simply too hard to grasp. I know a person with Tourette syndrome who has been subjected to two exorcisms. I have friends who were told that their Tourettic son's symptoms were the result of them not bringing the child to church often enough.

To me, this kind of ignorance is almost unbelievable. Have we not learned from the past? Is modern rational thought so boring that we would rather resort to ancient superstition? I am not condemning religion in any way but I am absolutely denouncing fanaticism and the lack of common sense.

Chapter Thirteen: The Freudians

With the exception of the horrors committed by the demon hunters of the Holy Inquisition, people with Tourette syndrome have suffered no greater disservice than the damage done to them by the Freudian psychoanalysts. While Sigmund Freud himself was a well-meaning, intelligent man, he did not have the resources to make the proper conclusions about many aspects of mental health. There was little information available in his day about the function of the brain. Knowledge of genetics was in its infancy at best. In Howard I. Kushner's excellent book, A Cursing Brain? The Histories of Tourette Syndrome, he explains in detail many of the bizarre theories that were argued until recently by the hard-core psychoanalysts.

Freud laid the groundwork; but his followers, including Sandor Ferenczi, took the practice of psychoanalysis to ridiculous levels, insisting that tics were the result of masturbatory urges and that any tic disorder was a "psychosexual" illness. Others claimed that tics commonly resulted from bad parenting, various aspects of hysteria, or childhood sexual abuse.

I think that Sigmund Freud and his followers should be respected for making pioneering attempts to dissect the the human psyche. However, the psychoanalysts who abused the respect given to them by their patients and by an answer-hungry public are to be blamed for the suffering of many.

Thanks to these therapists, countless loving, innocent parents were accused of abusive treatment of their children. Families were driven into severe conflict over the analysis and advice of these so-called experts and their irrational claims. The self-serving practitioners of this flimsy pseudo-science caused many Tourette syndrome patients and their family members to feel undeserved shame and guilt. The patient's tics often became much worse due to this added stress and some people actually developed psychological problems wholly induced by their treatment.

Much like the Inquisitioners, the views of the psychoanalysts about TS made sense only when one considers the information available at the time. Unfortunately Freudian-based psychology became a popular trend and more accurate medical explanations for tics were often pushed aside until the late 1960's. This shameful practice was in part a result of greedy psychologists who kept their good-paying customers in psychoanalytical therapy for years and years.

In the early to mid 1900's, Gilles de la Tourette syndrome was all but forgotten, buried deep beneath piles of books filled with "pop psychology" gibberish. Although doctors Arthur and Elaine Shiparo disproved the psychogenic theories about Tourette syndrome in the 1960's with their extensive haloperidol studies, many Freudians refused to accept the medical facts until the 1990's. Even now, some hold on to this ancient and misguided school of thought.

When I was a young college student, a psychologist told me that my tics were the result of my "screwing myself up." I brought them on myself by practicing poor thinking habits, nothing more.

Psychology, like medicine, has made great leaps and bounds. Today psychologists can offer helpful coping strategies that can truly benefit those struggling with TS and other disorders.

There are many psychological components associated with TS but it is nevertheless considered also a physical problem. In addition, many of the psychological disturbances found with the disorder result from the person's interaction with an often unforgiving and ignorant society. Even Sigmund Freud suggested that an organic component was possibly involved in tics. With today's successful treatments including deep brain stimulation implants, no rational person can argue total environmental causation over the electro-chemical nature of this disorder.

Modern psychological, pharmaceutical, and surgical therapies can all be effective in treating Tourette syndrome. Thankfully psychology and neurology are now working in concert, making them both much stronger disciplines.

Chapter Fourteen: Medication

Although cunning and intelligent, the Tourettic mind seems very foolish. It desires to tear down any hope of peaceful survival for its target yet it does not seem to realize that, without a lucid host, it too has no meaningful life. It seems to be unable to think past the act of sabotage.

It should have enough foresight to realize that the host may eventually seek medication for relief, especially if the symptoms get severe. With this knowledge, it seems that the Tourettic mind would learn to govern its actions. It could then carefully sustain an irritation level that allows the maximum amount of disruption without pushing its enemy to the point of launching a chemical counterattack.

When medication is used, no one usually gets by unscathed; both the ticcing mind and the conscious mind lose some ability to function. Many drugs kill the patient's creativity while others make it difficult for the person to sustain concentration. Without mental clarity, the Tourettic mind may end up somewhat as a harmless drifter. The host, however, does not have the option of letting the conscious mind live in a fog. If the person wishes to have any sort of social or professional life, he or she must eventually become somewhat coherent.

I have experienced this fog and I know how it hurts one's professional life. During the two years or so that I was prescribed Haldol, daily function became difficult because I would constantly fall asleep much like a narcoleptic person. In spite of these naps, I never felt rested. I was also unable to perform any task requiring even minimal concentration due to a perpetual state of confusion. This confusion brought on great frustration and depression.

Unfortunately, this was a period of time that I needed to be at my best. I was in a band had attracted the attention of a major record producer. He wanted to produce a record for us and we were excited about the project.

I spent two weeks in the studio, recording my guitar parts for a few minutes at a time and then falling in and out of sleep while the rest of the band continued recording. When I tried reducing the dosage of the drug, the tics would increase and I was embarrassed for this musical legend to see me twitching. Therefore I followed doctor's orders, took the prescribed dosage, and crawled away to sleep when I did not have to play. Oddly, my ability to play guitar was not severely diminished by the Haldol but my other cognitive skills were destroyed. It is still unbelievable to me that I actually got through the recording sessions in such a haze.

Because of my extremely dulled demeanor, the record producer probably thought that I was either a severe drug addict or that I was not all that interested in the project. Either way, I'm sure that I did not make the best impression. Nothing was ever said, but I have always wondered about it.

My handling of that situation was not very professional and I now wish that I could go back and explain my zombie-like state. However I cannot go back to those days and neither can the thousands of Tourettic students who were labeled "feebleminded" because of the side effects of medication. The medical treatment of TS sometimes does more harm mentally, physically, and socially, than the disorder's symptoms.

For this reason, most of us with Tourette syndrome spend our lives trying to decide if it is better to exist with or without medications. If the answer is with them, we struggle to find the correct drugs and the proper dosages that will allow us to somewhat function while at least reducing the symptoms.

I have no medical training and I do not want anyone to take anything that I say as medical advice. However, I do feel the need to speak out about the pharmaceutical treatment of this disorder carried out by some physicians. Most doctors now better understand Tourette syndrome but some are still using a 1980's approach to managing this disorder.

When I was growing up, Tourette syndrome was unknown to almost everyone. Tics were viewed by many as totally unacceptable acts. The few doctors that knew anything at all about TS were treating the disorder with only one goal in mind - stopping the tics. The medications used in those days would commonly put the patient in a stupor. However, if the drugs reduced the tics, they were considered good medicine.

Today, tics are somewhat better accepted by the public because more people now know about Tourette syndrome. We should take advantage of this increased awareness and physicians should stop routinely drugging children into stupors. Over-drugging was never truly necessary and it is even less needed today.

I have visited many schools to speak with students with TS, their teachers, and their parents. In about 80% of those meetings, the teachers have told me that the student's tics would not be that disruptive in class because the students could be educated about

Tourette syndrome. They said that the drugs, however, were causing significant harm by both preventing the student from learning and keeping the child from developing much-needed coping skills.

It is my non-professional opinion that more doctors should now consider treating the obsessive-compulsive symptoms and not even try to eliminate all of the tics. After all, it remains usually impossible to eliminate all tics with drugs while not rendering the patient at least somewhat cognitively blunted. This blunting does not commonly occur in the treatments used for obsessive-compulsive disorder (OCD). If the OCD-related symptoms are reduced, many of the triggering mechanisms behind the tics will consequently be reduced. To me, tics are merely the fallout; OCD is the bomb.

A physician can treat OCD with selective serotonin reuptake inhibitors (SSRI's) and other drugs that have much less side effects than many of the drugs used to directly address tic symptoms. This may not work well in every patient but it makes more sense than handing out neuroleptics to every child who displays tics. Although they have been much improved, neuroleptic drugs can sometimes rob a person all joy and clarity in his or her life.

Furthermore, I tend to agree with the researchers who feel that Tourette syndrome can actually be placed under the umbrella of obsessive-compulsive disorders. It simply makes sense to me. I think that the worst tics are reactions to intrusive thoughts and obsessions. The typical OCD-only sufferer has an obsession to complete a task perfectly. He or she will repeat attempts to carry out the task until it feels correctly accomplished. A Tourettic person will do the same; but this task to be completed is sometimes a tic.

Perhaps the best thing we can do it to treat the public, not the TS patient. It is true that awareness is now much better, but some people still seem judgmental about even a simple tic. The public needs to learn that Tourette syndrome affects a fairly large percentage of the population and that a person's ticcing should be looked at much in the way that one's sneezing or coughing is viewed. It is normal behavior, at least for a person with TS. In fact, tics should be even less alarming than sneezes because TS is never contagious.

Chapter Fifteen: No Rest for The Weary

I knew that it was late. I didn't need to see the clock. I had been lying in bed for hours attempting to settle into sleep. The acute awareness that I had to wake up early the next morning only added to my inability to sleep.

When I first lay in the bed, my mind had wandered at high speed, darting from one thought to another. These rapid thoughts finally started moving at a slower pace. As the calm began to overtake my body, at last I felt like I would actually drift into a pleasant sleep. I could feel the veil of relaxation gently fall over my mind as I started to drift.

Suddenly there was a severe jolt! I felt as if I had fallen and was awakened just before I hit the ground.

A horrific scene invaded my head with the speed and intensity of a gunshot. My heart jumped into instant tachycardia as I felt an odd rush of both terror and sorrow. The hideous vision was of the attempted rescue of a mutilated car crash victim. I could smell the rich stench of transmission fluid, gasoline, and anti-freeze as it belched from the twisted metal that once was a sedan.

There was another smell in the mix that I could not identify. It was possibly the smell of medical chemicals, like one finds in a hospital. My mind rapidly scanned through a number of memories in an effort to recognize the chemical scent. Unfortunately the odor was too vague. Perhaps it was the sickening, sweet smell of flowers. Many flowers emit a suffocating, repulsive odor to me and this stench was similar. In spite of the odor's mysterious source, I knew exactly what it represented to me; death. I was instantly distraught.

As the flash of red and blue lights from the emergency vehicles interrupted the dark night, I could see two paramedics lifting a woman's limp, bloody body from the driver's seat. Smoke was rising from an area to the left of my view, indicating the possibility of more wreckage. I felt that something even worse than I was observing had happened and more carnage lay just out of my view. I could feel the panic increasing.

As I struggled to recognize the vehicle and the surroundings, I scrambled to identify any of the people in the vision. It was impossible to identify them yet I knew that I was somehow closely connected to the crash victim. I could feel the agony of deep sadness and the permanence of death as I quickly digested this horror before me. The scene was somewhat confusing, filled with uncertainty; but it was both intense and personal. I was also fully convinced that I was awake and that this was no nightmare.

My horror was quickly interrupted as I felt my fist repeatedly hitting myself on the side of the leg. My attention suddenly shifted to this action and I realized that this physical attack had been triggered by an undeniable urge that lasted only a nanosecond. The urge had quickly crept up during my distraction with the vision.

A part of my brain seemed to have been aware of the approaching sensation yet it had too little time to warn the rest of the mind. By the time the physical tic began, the premonitory urge was a recent memory.

In tics like this one, the intrusive thought and the rapid urge work in concert. In other instances, the precursory sensation may build up over a few seconds and fail to sneak in under the cloak of a mental diversion; the urge can be felt as it expands.

After a few blows to my leg, the pain in the bruised area seemed to indicate that the tic been carried out correctly. I was now wide awake, upset, angry with myself, and I could feel the trauma of the minor injury. Therefore the Tourettic mind was totally satisfied with its victory and the tic process was completed.

The urge quickly dissipated. It fled into the shadows to prepare for another ambush and awaited its next orders from its commander, the Tourettic mind. I was so relieved that the horrible crash scene was erased from my mind that the pain did not bother me at all.

More importantly, and in spite of all rationale, I believed that this violent tic was some sort of magical device for preventing such a terrible crash from happening. It was as if the crash scene was a premonition and the performance of the tic was the only way to alter the inevitable, to change fate.

Another odd theory was that I had experienced a form of remote viewing of an actual event in real time. If that were the case, the tic seemed to render the crash scene unreal. It made it go away as if it were a bad dream and it never really happened. Had I not gone through with the tic, the vision would be valid, not a dream. It would likely be soon to occur, have occurred, or be in the process of happening.

The entire crash-scene vision lasted only a few seconds and it was abruptly stopped when the tic was completed. Unfortunately the next step was to obsess about what the whole thing meant. I also had to be concerned about whether my "preventative tic cure" was effective. Did it stop the reality of this horrible occurrence? If the accident were to be in the future, did my tic alter the course of destiny?

Although I knew that these ideas about premonitions and preventative measures were ridiculous, my Tourettic mind was telling me that all of it was perfectly logical. After all, the Tourettic world has its own logic, alternate to that of the everyday rationale. This is the OCD-tic combination and the magical thinking that it causes.

I can never truly believe such magical thinking can be valid in the real world. Rational thought and common sense tell me that. But what if it is right and my common sense is wrong? What if I fail to carry out the tic and something horrible happens? I don't want that blood on my hands and I don't want someone to die if I can stop it. It doesn't matter if my rational mind can comprehend these absurd rules or not; the risk of challenging them is simply too great.

The aftershock of this vision consisted of a ten-minute sequence of related obsessive thoughts and mild vocal and motor tics. I tapped my bruised leg a few times and grunted quietly as I tried to wind down from the excitement and perform even more preventative measures in the form of mild motor tics, just to be safe. This is much like the behavior of a superstitious person. Perhaps OCD and some tics can be described as "hyper-superstitious" behaviors.

During a bad night, this pattern repeats itself for hours. Other nights may pass with less disturbing visions, pleasant visions, or no major interruptions at all. Often I will go to sleep rather quickly, sleep for an hour or so, and then wake up to begin a cycle of intrusive thoughts.

Most nightly disturbances involve my starting to nod off and then being stunned by the intruder with just enough shock to wake me up. The horrible visions often never fully develop but rather quickly flash in and out as if they were one or two upsetting frames edited into an otherwise pleasant movie. This game goes on for hours, taking about one to two minutes per cycle. The anti-self's purpose in these situations is to ruin my chance of getting some much-needed rest.

No matter how the sequence of events unfolds, a good night's sleep is impossible for me without the use of medication. In an effort to avoid being a total sleeping pill addict, I sometimes suffer through the night with no medication and try to get through the next day without adequate sleep. The older I get, the harder it is to function without sleep. I have also tried every non-drug routine that I know of to encourage sleep.

Those who do not experience obsessions and tics often fail to understand why people with TS cannot simply relax. For many people, reading a good book or listening to soft music is relaxing.

Reading does not work for me because I inevitably run across words or phrases in the text that set off obsessions, compulsions, and tics. The same potential problem exists when watching television. However reading, for some unknown reason, is much more likely to trigger these negative responses.

Listening to soft music is absolutely out of the question for me if I ever want to sleep. Being a musician, I cannot help but analyze every chord progression, every melody, and every rhythm pattern involved. I cannot stop my mind from trying to play along with the music.

I was given a CD of music designed to induce relaxation and sleep. I gave it a try but it kept me awake all night. I could not use it at all. Furthermore, it is difficult for me to imagine anyone being able to go to sleep with any type of music playing.

Silence is also intolerable when I am trying to rest. It is an enemy to me. Not only is extreme quietness bothersome in itself, there is no masking of outside sounds. Every dog's bark and every street sound is a rude interruption. I become acutely aware of any significant variation in the ambient decibel level and jump into total alertness with every abrupt change.

The best sound environment for my sleep is a constant, hypnotic drone such as a fan motor or a noise-generating machine. I have a noise generator and it helps a bit if I keep it on the "white noise" setting, which is a constant flow of sound. However, from time to time, my ear picks up a galloping sound of repeated rhythms in the machine's sound generator. At that point, I start to mentally play along with this rhythm and the device becomes simply another sleep-killing irritant.

The bottom line is that, every night feels to me as if my Tourettic brain is trying to sabotage my sleep. It knows to launch its strongest attacks on the nights that I desperately need a good night's rest. That way, the next day will find me tired, my defenses will be down, and I will consequently be more susceptible to its daytime attacks.

This pattern has continued for so long that, if asked to describe the feeling that a good night's natural sleep provides, I cannot offer an answer. Not at all. I may have occasionally slept well as a child but I can no longer remember how it felt. Through the years, I have figured out ways to work around these sleeping problems and I generally get enough rest. Therefore my irregular sleep is not that hard for me to deal with.

Chapter Sixteen: Summary

The experiences I have shared in this book may seem odd to many people. However, according to my friends who have Tourette syndrome, my stories are pretty typical. The circumstances may be somewhat different but the overall sensations and disruptive patterns are usually the same. Many of us sense that we are battling an internal saboteur that is determined to destroy every moment of peaceful enjoyment we may have. To many observers, this seems sad and disturbing. To most people with TS, it is a way of life and most of us handle it pretty well.

A few years ago I created a pilot for a documentary film that I had hoped to someday make about Tourette syndrome. I wanted to use the power of multimedia to give non-Tourettic people a better understanding of how tics, obsessions, and other symptoms may feel to the affected person.

I conducted a showing of this 30-minute movie pilot at a meeting of people who work with Tourette syndrome patients. I thought that it would help them to understand what some of their patients with TS could be going through. After all, the film was aimed at adults and it was not made to be viewed by children. The movie contained nothing offensive; it was simply an attempt to present both tic and obsession experiences in a movie form.

As soon as the film started, this so-called educated audience became upset. As I demonstrated on the screen what an intrusive thought and an obsession-driven tic could feel like, they recoiled in horror as if I had shown them an atrocity film. When the movie ended, one of them said, "Rick, you cannot show this to parents; they cannot handle it." These support people and caregivers were almost angry with me for exposing the most upsetting experiences that someone with Tourette syndrome may endure.

There were also perhaps six audience members there who had TS. They all calmly said, "Yes, that is pretty-much the way it is" and they were not in the least bit disturbed by my film.

This led me to the conclusion that many people underestimate the inner strength possessed by those of us with Tourette syndrome. They forget that, most of us have grown up relentlessly fighting this saboteur inside us and it has become somewhat routine for us to do so. We still have to face the same outward challenges as anyone else but we do so while also carrying out a full-scale war inside our minds.

Therefore, as adults we don't need pity or special treatment. As children, we don't need adults overreacting to our tics. We should also never feel pressured to be drugged out of our minds just so those around us will not be disturbed by our tics.

We are strong enough to fight this internal battle. All we need to win it is a little compassion, a bit of understanding, and the simple respect of others.